Handbook of Poultry Diseases

Handbook of Poultry Diseases

Naresh Mahajan

RANDOM

RANDOM PUBLICATIONS

NEW DELHI (INDIA)

Handbook of Poultry Diseases

ISBN 978-93-5111-381-2

Published in 2014 in India by

RANDOM PUBLICATIONS

4376-A/4B, Gali Murari Lal, Ansari Road
New Delhi-110 002
Phone : +91-11-43580356, +91-11-23289044
e-mail: randomexports@gmail.com, sales@randompublications.com, info@randompublications.com

Reprinted 2026

Type Setting by: Friends Media, Delhi-110089

Preface

If a poultry producer is to reap all the benefits from the investment of money, time and labour, he must maintain a healthy and parasite-free flock. The health status of his birds is important to the producer whether he maintains a backyard flock or commercial poultry flock. Constant monitoring of the flock is the first step in detecting diseases. Identification of the cause and appropriate treatment can prevent a minor condition from becoming an unprofitable enterprise. There are many common and important diseases which can affect the respiratory system of poultry. Poultry refers to birds that people keep for their use and generally includes the chicken, turkey, duck, goose, quail, pheasant, pigeon, guinea fowl, pea fowl, ostrich, emu and rhea. Due to modern systems of management, usually with high poultry densities, these diseases are able to readily spread. The symptoms vary according to the age of the affected birds. The first symptoms usually observed in young birds are sneezing, gasping and often droopiness. It is in this stage of the disease that the manifestations rather closely resemble those of infections bronchitis. Within a short time after appearance of respiratory symptoms, deaths occur in a flock in quick succession and in increasing numbers from day to day. Among growing birds and in adult sudden deaths occur in a dew instances, and are followed by a number of birds showing respiratory symptoms. The affected birds are full and depressed with ruffled feathers. These symptoms are accompanied by diarrhea, characterize by the passing of a watery stool with an offensive smell. There is profuse salivation. The saliva often accumulates in the mouth and obstructs respiration, which results in the production of gurgling disused birds may be soft - shelled and deformed. In turkeys the disease runs a very mild course. In adults, in particular, it may pass unnoticed except for some dullness, loss of appetite and other minor symptoms.

At present there is no effective treatment of any value. Proper housing and general good care are indicated in an effort to shorten the duration and severity of the infection. An early recognition of the disease and application of struck sanitary measures are of great value in the control of the disease. Some important measures for its prevention are ; slaughtering of all apparently ailing birds, segregating of in - contact in group of 10 to 15 each; removal of all infective materials such as droppings, residues of poultry cleanliness ; and provision of separate attendants for each group of birds. The poultry farm should be at a distance from place of traffic. All newly purchased birds should be kept in segregation from not less than 10 days before taking them into the farm. The poultry runs should be ploughed from time to time and lime applied thereon as a general disinfectant. As far as possible the pens and runs should be made inaccessible to free - flying birds by providing a barrier of wire - netting. Control of Ranikhet diseases can be effect with judicious application of sanitary and vaccination measures. The possibility of entry and spread of infection is considerably reduced through the maintenance of flock on deep little system and stopping all unauthorized entries, even of human beings, into the battery brooders. Disposal of fowl carcasses by burning or deep brutal to reduce the scope of carrion-eating birds like crows, kites and vultures perching near fowl pens or poultry farms helps to resume the hazards of this infection. Two types of vaccines are available in India, one for the adult birds an another for younger birds or body chicks. The virus strain for Ranikeht disease vaccine used for adult birds age over weeks was evolved at the Indian veterinary Research Institute. The vaccine consists of freeze - dried virus grown in chick embryos. Vaccination of birds 6 weeks old and above confers immunity for 1 to 3 years. Care should be taken to vaccinate bird's not carrying heavy coccidian infection. Birds with heavy worm infection or coccidiosis are not protected even with a good vaccine. There are sometimes complications side reactions following vaccination. There are sometimes complications side reactions following vaccination with 'Mukteswar' strain of Ranikhet disease vaccine. These consist on inco-ordination of limbs and sometimes paralysis in 1 to 3 per cent to the vaccinated birds. The reactions may become more acute if the birds are affected with roundworms, coccidiosis or are weak on account of malnutrition.

This book explores the latest advances in the field of this subject. The subject matter, both as regards the arrangement of chapters as well as contents is designed to meet the requirement of the students in several Universities.

I thank all members of my team who have helped in the preparation of the book. My special thanks go to "Random Publication" who have published the book.

—Naresh Mahajan

Contents

1

Bacterial Diseases in Poultry

Escherichia Coli Infections

Omphalitis (Navel Infection)

It is characterised with reddening and tissue oedema in the umbilical region. Escherichia coli infections are widely distributed among poultry of all ages and categories. They are primarily related to poor hygienic conditions, neglected technological requirements or to respiratory and immuno-suppressive diseases. A common sequel of navel infections is local or diffuse peritonitis. When the amount of egg white is bigger (in larger eggs), it impedes the absorption during hatching, resulting in subcutaneous jelly-like oedemas that are an excellent media for the development of E. coli infections.

The delayed absorption of the yolk sac is a prerequisite for E. coli infections and peritonitis. The most commonly identified E. coli serotypes are: 01:K1 (L); 02:K1 (L) and 078:K80 (B).

At a later stage of the infection, the yolk content is a cause of putrefactive necrotic processes in the peritoneal cavity. The abdomen is bloated. The entire abdominal wall is affected by a moist gangrene (maceration).

Salpingitis (Inflammation of the Oviduct)

Salpingites due to E.coli infections could be also observed in growing birds. The oviduct is dilated, with thinned wall and filled with caseous exudate all along its length. Salpingites are among the commonest causes for death in layer hens. E.coli penetrates from the cloaca via an ascendant route. Predisposing factors are the intense egg laying and the associated estrogen activity. In older cases, the caseous masses in the oviduct have a lamellar structure. E.coli organisms are usually found in excreta because of their presence in avian and mammalian intestine, the birds are constantly at risk of infection through contaminated water, dust, faeces and environment.

Retained yolks among the caseous masses in the oviduct. In some cases, when the systemic resistance is lower, places, contaminated with E.coli, such as intestine, genital tract or nasal passages, could be latent sources of infection. Compression and compactedness of caseous necrotic masses after losing a part of their water content in the oviduct of a layer

Egg yolk peritonitis in a layer hen consequently to E.coli salpingitis. The chickens could be hatched with a latent infection, when E.coli is present in ovaries and the oviduct. In these instances, the infection could turn into an overt infection under the influence of some stress factors or lesions. (Inflamation of the ovary) consequently to a salpingitis due to ascendant E.coli infection. Cystic degeneration of ovarian follicles following an E.coli opphoritis.

Cellulitis

Cellulitis Inflammation of the subcutaneous tissue that affects also the overlying skin. It predominates in broilers and is detected mainly in slaughter-houses. Macroscopically, the lesions are with a yellowish-brown colour. Cellulitis effected areas are mostly in the region of the back and the thighs. In some cases, the lesions are slightly prominating over the adjacent healthy skin. In the subcutaneous tissue, thick fibrinous plaques are often found out. In some cases with adult birds, in the region of the head, subcutaneous masses of thick serofibrinous exudate resulting from a local E. coli infection could be detected.

Enterocolitis

Enterotoxigenic E. coli that produce toxins, cause the secretion and retention of fluids in some intestinal loops and especially in the caeca. Clinically, diarrhoea and dehydration are observed. The intestines are pale and distended, particularly the caeca that are overfilled with fluid containing many gas bubbles.

Neonatal E.coli Septicaemia: Chickens in the first 24 - 48 h after hatching are affected. The death rate during the first ten days is higher and could reach 5-6%. The yolk sac is unabsorbed. The spleen is enlarged. Some days later, the typical serofibrinous polyserositis lesions, affecting the peritoneum, the pericardium, the air sacs and the liver capsule are manifested.

Acute E.coli septicaemia in layer hens. Clinically and morphologically, the acute E.coli septicaemia could resemble fowl cholera or fowl typhoid. It is encountered in both young and sexually mature birds. The stress in the beginning of egg-laying is considered as an important predisposing factor. The parenchymal organs are enlarged and hyperaemic. Sometimes, the liver has a greenish colour and is mottled with multiple small necrotic foci. Also, pericarditis, peritonitis and petechial haemorrhages on serous coats are present.

E.coli septicaemia of a respiratory origin. In such cases, the respiratory mucosa damaged by infectious and non-infections agents (ND viruses including vaccinal strains, IB, TRT, mycoplasmae, high ammonia levels) is the entrance door of the E.coli infection. The lesions are principally observed in the respiratory tract (trachea, lungs and air sacs), but some adjacent serous coats (pericardium, peritoneum) are also affected and thus, the picture of a typical serofibrinous polyserositis is produced.

E.coli septicaemia secondary to enteritis. It is most commonly encountered in turkeys. The intestinal mucosa, damaged by the haemorrhagic enteritis virus, is the entrance door of E. coli infection. The most typical lesions are the marked enlargement, hyperaemia, haemorrhages and necrosis of the liver and the spleen.

Panophthalmitis (Inflammation of All Tissues of the Eyeball)

Generally, it develops secondary to E.coli septicaemia and is usually unilateral. Arthritis, Osteomyelitis and Osteonecrosis (inflammation of joints, bone marrow and bone necrosis, respectively). The lesions are a common sequel to E. coli septicaemia. Clinically, lameness, prolonged lying down, dehydration and retarded growth rate are observed. The coxofemoral joints, the femur and tibiotarsal joints are most commonly affected. The bacteria colonise the physes of growing bones and provoke an inflammatory response that is further causing osteomyelitis. Pathoanatomically, fractures of the femoral head are usually discovered. Some cases of E.coli septicaemia, accumulation of exudate between the superficial and deep pectoral muscles is observed. The lesions that develop in the articular spaces of thoracolumbar vertebrae result in spondylitis (spondylosis) and after that, in progressive paresis and paralysis.

Coligranuloma (Hjarre's Disease)

It is characterised by multiple granulomas in the intestinal tract, the mesentery and the liver, but not in spleen. The lesions are similar to these observed in tuberculosis.

Bursitis Sternalis (Inflammation of the Sternal Bursa)

The bursa is enlarged in a various extent and filled with inflammatory exudate. The diagnosis of coli-infections is based on isolation and typisation of pathogenic E.coli serotypes. Many other bacteria (salmonellae, pasteurellae, staphylococci, etc.), viruses, chlamydiae and mycoplasmae should be excluded as possible aetiological agents. The prevention should aim at minimising the probability of faecal contamination of eggs. This implies the maintenance of clean nests, discarding floor eggs and removal of eggs that are cracked or contaminated with faeces. Breeder eggs should be fumigated or disinfected in the farm prior to their transportation in the storage premise. The treatment is effective if initiated soon after testing the antibacterial sensitivity of isolates.

Pullorum disease is an acute systemic disease in chickens and turkey poults. The infection is transmitted with eggs and is commonly characterised by a white diarrhoea and high death rate, whereas adult birds are asymptomatic earners. The morbidity and the mortality rates increase about the 7th - 10th day after hatching. The affected chickens appear somnolent, depressed and their growth is retarded. The feathers around the vent in many chickens is stained with diarrhoeic faeces or pasted with dry faeces.

The oedema of tibiotarsal joints is a frequent associated sign. Pullorum disease is widely distributed among all age groups of chickens and turkeys. The highest losses are in birds under the age of 4 weeks.

The aetiological agent is S.pullorum, a non-motile Gram-negative microorganism. S. pullorum is very resistant under moderate climatic conditions and could survive for months. It could be killed by fumigation with formaldehyde of breeder eggs in the hatchery. Typical for this form are the greyish-whitish nodes in one or some of the following places: heart (39), lungs, liver, gizzard walls (40) and intestines, the peritoneum.

Sometimes, greyish-whitish milliary necroses are found out in the liver. S.pullorum is transmitted by infected eggs of layer hens that are carriers. Many hatched infected chickens spread the microorganism by a horizontal route to other birds via the gastro-intestinal and the urinary tracts. Adult carrier birds also spread the agent through their excreta. Ureters are often filled with urates. For confirmation of the diagnosis, pullorum should be isolated and typed. Pullorum disease must be differentiated from other salmonelloses, E.coli infections, Aspergillus that produces similar pulmonary lesions. Staphylococcus aureus, causing arthrites, etc. Sometimes, the pulmonary nodes resemble the tumours in Marek's disease.

Acute Fowl Typhoid

The outbreaks usually begin with a sharp decline in forage consumption and egg production. The fertilization and hatchability rates are considerably reduced. Diarrhoea appears. The death rate in acute fowl typhoid is high and varies between 10% and 90%. About 1/3 of chickens hatched from eggs from typhoid-infected flocks die. A characteristic lesion for acute fowl typhoid in adult birds is the enlarged and bronze greenish tint of liver.

In some instances, the enlarged liver is mottled with multiple milliary necroses. The outbreaks are observed primarily in hens and turkeys, but the disease is sometimes encountered in other domestic or wild fowl. In other cases, the size of liver necroses varies from milliary to spots with a diameter of 1-2 cm. Unlike pullorum disease, fowl typhoid is lasting for months. Acute fowl typhoid. The spleen is 2-3 times bigger, sometimes with greyish-whitish nodules prominating on the surface, representing hyperplasic follicles. Often, enteritis, especially of the anterior part of small intestine, sometimes with ulcerations, is present. The aetiological agent is Salmonella gallinarum. This organism usually shares common antigens with 5. pullorum and the two microorganisms often give a cross-agglutination reaction.

More rarely, myocardial necroses due to Salmonella toxins are detected. The transmission of the infection by contaminated eggs is especially important. Moreover, the transmission of 5. gallinarum occurs mainly among growing or productive flocks and the death rate among adult birds is higher. The lungs acquire a characteristic brown colour. Here, necroses and, following their organisation, „sarcoma-like nodules"

Chronic Fowl Typhoid

The lesions are primarily in the gonads. The ovaries are affected by inflammatory and degenerative changes. Frequently, affected follicles are deformed and appear like thick pendulating masses. Fowl typhoid should be differentiated from other salmonelloses, E. coli infections, Pasteurella spp. infections, etc. If breeder flocks are proved to be carriers of the infection, their eggs should not be used for breeding.

Sometimes, the going out of yolk from degenerated follicles results in fibrinous adhesive peritonitis. Taking into consideration that chemotherapy does not eliminate the carriership, the treatment of poultry infected with fowl typhoid or pullorum disease is not justified and is never recommended.

Riemerella Anatipestifer Infections

Riemerella anatipestifer (RA) infection is a contagious disease in domestic ducks, turkeys and other fowl. It is encountered as acute or chronic septicaemia and is characterised with serous fibrinous polyserosites. The respiratory tract could be also affected. The ducklings at the age of 18 weeks are especially susceptible. Clinically, sneesing, cough, trembling of the head and neck, ataxia and greenish diarrhoea could be present.

The most characteristic gross lesion is the deposit of fibrinous exudate on the pericardium, the liver capsule or air sacs. The chronic lesions affect the skin and the joints. Although a tentative diagnosis could be made on the basis of observed clinical symptoms and lesions, it is confirmed upon the isolation and identification of RA. The RA infection should be distinguished from septicaemiae due to P. multocida, E.coli, Salmonella, etc. The treatment with antibiotics (Flumequine) and sulphonamides IJrimetoprim, Sulphadiazine) has a varying success.

Staphylococcus aureus infections are commonly seen in poultry. Usually, bones, tendon sheaths and joints, particularly coxofemoral and tibiotarsal joints are affected. The clinical signs include unilateral or bilateral lameness, reluctance to move and lying down. When the tibiotarsal joints are affected, swellings, fever and sometimes necroses of overlying tissues and purulent exudation are observed. As a sequel to septicaemia, osteomyelitis could occur. The lesions are usually detected in the region of the proximal femur, where inflammatory necrotic foci in the bone marrow and partial or complete fracture of the femoral head are observed.

The outcomes of skin injuries are the appearance of cellulites, characterised by extensive purulent inflammation of the subcutaneous tissue. Most staphylococcal isolates are identified as Staphylococcus aureus, Gram-positive cocci, arranged in clusters. Plantar abscesses are a kind of local purulent inflammation. They are located on the plantar surface or adjacently to it and appear following puncture wounds. Staphylococci are moderately resistant to common disinfectants. Chlorine-containing disinfectants are efficient in the absence of organic matter.

Other local forms of staphylococcosis are sternal bursites. The sternal bursa is enlarged at a various extent because of gathering of purulent substance. Sometimes, the covering skin is necrotised. The microorganism is widely distributed in the environment and mainly on the skin. Most of Staphylococcus oureus-induced lesions are associated with skin injuries, debeaking, finger cuts. All categories of birds are affected.

Toxicogenic strains are able to induce food intoxications. In a septicaemic staphylococcal infection, hyperaemia, enlargement and various-degree coagulation necroses in the liver or the spleen are observed. As staphylococci are ubiquitous, their presence could not be prevented. The measures should be directed towards minimising the possibilities for traumas of skin, respiratory and intestinal mucosa.

A typical secondary infection is staphylococcal gangrenous dermatitis that is usually resulting from immunosuppressive infections (IBD, CIA). Affected skin areas are dark red to blue-greenish, moist and sharply defined from the adjacent healthy skin. Histologically, staphylococcal lesions consist of necroses, bacterial colonisation and heterophilic infiltration. The isolation and identification of Staphylococcus spp. confirms the diagnosis. From a differential diagnostical point of view, infections with E. coli, S. gallinarum, P. multocida, M. synoviae and reoviruses should be considered.

Streptococcosis in poultry is encountered as acute septicaemia or chronic infections with a death rate between 0.5 and 50%. The agent that is most commonly associated with several pathological conditions in poultry is 5. zooepidemicus. The lesions in chronic streptococcal infections include arthrites, tenosynovites, myocardites and valve endocardites. Endocardites affect predominantly mitral valves and less frequently, aortal and tricuspid valves.

Other lesions related to endocardites, are heart, liver and spleen infarctions. Liver infarctions are usually peripherally located on margins, have a pale creamy colour and are sharply demarcated. The penetration of the infection occurs mainly via the oral or aerogenic route, but could also enter through the injured skin, especially in battery cage layer hens. The differential diagnosis includes other bacterial septicaemias as staphylococcosis, fowl cholera, E. coli infections, etc.

Mycoplasma

MG is characterised by respiratory symptoms and a prolonged course of the disease. Particularly susceptible are hens and turkeys at all ages. The aetiological agent is M. gallisepticum. In many cases however, the pathogenicity of the microorganism is enhanced because of its association with any or some of the following agents: E. coli, P. multocida, H. paragallinarum and IB or ND viruses. The most characteristic signs in adult flocks are tracheal rales, nasal discharge, coughing, decreased egg production. Most outbreaks are in broiler chickens older than 4 weeks. The course of the disease is more severe during the winter and in cases of associated infections. Often, conjunctivrtes, facial skin oedema and profuse tear secretion could be observed.

In turkeys, unilateral or bilateral swelling of periorbital sinuses, nasal discharge and conjunctivitis are observed. The inflammatory exudate is commonly fibrinous and is detected as diffuse accumulation after removal of overlying skin.

Most common gross finding is aerosaccultes, the air sacs being filled with fibrinous caseous exudate. The majority of routine chemical disinfectants are effective against M.gallisepticum that rarely survives longer than a few days away from the host. The birds could carry the microorganism and be asymptomatic until the disease is triggered by stress factors such as change of the premise, the diet or weather, vaccinations against or infections with IB or ND, increased levels of dust or ammonia.

In older cases, the content of air sacs is dense and compact. A vertical transmission is done through the eggs of some unapparent carriers. The infected progeny transmits the agent horizontally via airborne route, by coughing or contaminated forage, water and environment. Serofibrinous pneumonias, usually bilateral, are a frequent finding.

Often, the inflammation involves the adjacent serous coats and thus, fibrinous polyserosites occur. Sinusites are relatively rarely observed in hens. The positive agglutination tests of sera in several birds from the flock confirm the diagnosis. MG should be distinguished from other respiratory diseases in poultry. Pulmonary and air sacs lesions could be mistaken with similar findings in E. coli septicaemia or aspergillus's. In turkeys, P. multocida pneumonia should also be considered.

Mycoplasma synoviae (MS) infections could progress as either acute or a chronic systemic disease with symptoms of arthritis, synovitis and bursitis especially in hens and turkeys. The earliest signs are lameness, lying down and retarded growth. Often, oedemas of tibiotarsal joints and the drumstick are observed. The morbidity and death rates are moderate, under 10%. Young chickens at the age of 4-12 weeks and turkey poults at the age of 10-12 weeks are susceptible. Synovites are encountered all year round, but are prevalent during cold humid seasons or when the litter is wet.

Affected birds get progressively exhausted. When the joints and tendon sheaths are open, a serofibrinous exudate is most commonly observed. The aetiological agent is M. synoviae. The microorganism shows a certain tropism to synovial structures as joints and tendon sheaths. An important route of dissemination of the agent is the transovarial transmission. The distribution by a horizontal route via the respiratory tract is also possible. The commonly used means of diagnostics is ELISA. MS infections should be differentiated from staphylococcal infections, reoviral arthritis and RGT.

Necrotic Enteritis

Necrotic enteritis (NE) is an acute Clostridium infection characterised by severe necroses of intestinal mucosa. The disease begins suddenly, with a sharp increase in death rate. A strong dehydration is observed. The skin is sticked on or adhered to body musculature and is hardly removed. Chickens at the age of 25 weeks are usually affected, NE is also encountered in hens particularly near the period of the beginning of egg laying or peak egg laying, most commonly associated with coccidiosis. In acute cases, marked congestion of liver, responsible for its dark red to black appearance, is present.

The aetiological agent is Clostridium perfringens, mainly from type A and more rarely from type C. The produced a and p" toxins, from C. perfringens type A and type C respectively, are responsible for the necrosis of intestinal mucosa. The small intestine is often distended with gases and the necrotic mucosa is visible through the wall. CI. perfringens is ubiquitous and normally reside into the intestinal tract. The alterations are particularly in the jejunum and the ileum because of their higher pH and the lower oxygen content in these areas. Sometimes, haemorrhages are seen through the intestinal wall. The intestinal lumen is filled with brownish watery content, mixed with gas bubbles. Necrotic mucosa acquires a greyish-creamy or greenish appearance. Sometimes the mucosa has a flannelette blanket-like appearance.

In some cases, the mucosa has a linear pattern similar to the bark of a tree. The predisposing factors are injuries of intestinal mucosa by various Eimeria species, migration of ascarids, immunodeficiency states due to CIA, IBD, MD, high content of wheat or fish meal in the diet.

In cases when NE is associated with small intestinal coccidioses, multiple petechial haemorrhages could be perceived through the wall in different areas along the small intestine. Throughout the simultaneous occurrence of NE and coccidioses, the content of the lumen is bloody, mixed with necrotic detritus and gas bubbles. The diagnosis is based on the distinctive gross lesions. When necessary, a histological investigation is performed or attempts for isolation of the causative agent. NE should be distinguished from ulcerative enteritis and some small intestinal coccidioses. The control should be aimed at predisposing factors. An appropriate medication of feeds is recommended. A good effect is obtained with oxytetra-cycline dihydrate (OTC 50% premix). NE could be effectively treated with doxy-cycline hydrochloride, amoxicillin, etc.

Cholangiohepatitis (CAH) in broiler chickens is characterised by inflammatory proliferative and dystrophic necrobiotic alterations in bile ducts and the liver parenchyma. Usually, no clinical signs are observed. The increased daily mortality is insignificant, although in some chickens, a retarded growth and dehydration could be present. Pathoanatomically, the liver is enlarged and with paler yellow colour. In some cases, its surface has a characteristic acinous appearance and in others is mottled with multiple small greyish-white or greenish foci.

The walls of the gall bladder are thickened, sometimes up to 5-6 cm, and opaque. The state is detected in the last phase of the fattening period or in the slaughterhouse. It is possible to observe CHA as an independent disease or associated with necrotic enteritis. A transverse cross section through the thickened gall bladder wall. Clostridium perfringens is the aetiological agent. CAH is experimentally reproduced in broilers by ligation of bile ducts and inoculation with CI. perfringens. The gall bladder is filled with a thick bile secretion or a dense matter with a creamy colour.

In some chickens, the subcutaneous fat and the body fat have an icteric tint. Histologically, the liver lesions are detected at a various stage of development. In the majority of cases, proliferative changes in bile ducts are observed. The overgrown bile ducts form granulomatous structures, surrounded by fine reticular fibres. Centrally, in some granulomas, either initial or advanced degree of necrosis and weak to moderate granulocytic infiltration are observed.

In many bile ducts, a biliary stasis is present and within some, among the stagnated secretion, a huge amount of microorganisms are detected. Pericanalicularly, coagulation necroses are frequently noticed. Among some of these foci, hyalinisation of the necrotic masses and single microorganisms are present. Around the necroses, a belt of macrophages, lymphocytes and granulocytes is formed.

Many Gram-positive bacteria are detected among granulomas, in bile ducts' lumen and into the gall bladder, often accompanied by inflammatory lesions.

The mucosa of bile ducts and the gall bladder is frequently necrotised and the wall is thickened because of connective tissue growth. The diagnosis is based on the characteristic gross and microscopic lesions. The prevention is similar to that in NE.

Ulcerative enteritis (UE) is characterised by inflammatory ulcerative and necrotic changes in intestinal mucosa and dystrophic necrobiotic lesions of the liver and the spleen. The clinical signs include a general malaise, ruffled feathers, diarrhoea and anaemia. In many instances, the disease begins with a sudden death. Pathoanatomically, deep button¬like ulcers are observed, mainly in caeca and less frequently, in some parts of the small intestine, usually visible through the wall.

UE is a problem in all world regions with extensive poultry breeding. Young birds are infected more frequently although the disease is also common among adult quails. The early lesions appear like yellowish foci with haemorrhagic boundaries that could be seen from both the serous and the mucosal surfaces.

The aetiological agent is Clostridium colinum, a spore-forming organism, highly resistant to chemical agents and physical alterations. The intestinal content is often mixed with blood.

In older and larger ulcers, the haemorrhagic zones tend to disappear. The ulcers could have an irregular round or elongated shape and are covered by large necrotic diphtheritic membranes.

Frequently, adhesive peritonitis due to inflammatory involvement of adjacent serous coats is observed. Numerous domestic and wild birds (chickens, quails, turkeys, rock partridges, geese, partridges, etc.) are susceptible. The chickens and the quails are the most vulnerable between 4 and 12 weeks of age where as turkeys between 3 and 8 weeks of age. Liver, a variety of dystrophic changes and necroses with different size and shape are detected. Necrotic foci in some cases are milliary. Sometimes, liver necroses reach 1 - 2 cm. in diameter and are surrounded by a haemorrhagic zone.

In some instances, necrotic foci involve large parts of the liver and are infiltrated by haemorrhages. The aetiological agent is distributed with the excreta of acutely ill and recovered birds and persists in soil for many months. The incubation period is 1 - 3 days. The death rate in chickens varies from 2% to 10% and in quails reaches 100%. The outbreaks of UE in chickens are often associated with or come after coccidioses, CIA, IBD or stress conditions. Most commonly, necroses are distinguished on the background of a marked parenchymatous dystrophy, affecting partially or totally (uni- or bilaterally) the liver. The spleen could be enlarged, hyperaemic, haemorrhagic and sometimes, with necroses. The diagnosis is based on the typical gross lesions. When needed, imprint preparations are made, a histological study is performed or attempts for isolation and identification of the aetiological agent are made. UE should be differentiated from NE, coccidiosis and histomonosis (typhlo-hepatitis).

Some cases, haemorrhages with various intensities are detected in the mucous coat of the gizzard. Prevention -separate housing of the different age groups of birds, avoiding the contact with other avian species. The pre-medication of forages with some antibiotics and their rotation would prevent the reproduction of CI. colinum. A good effect is achieved with oxytetra-cycline dihydrate (OTC 50% premix). UE could be effectively treated with doxycycline hydrochloride, amoxycillin, etc.

Gangrenous Dermatitis

The gangrenous dermatitis (GD) is a disease affecting fattening and growing birds, characterised with necrotisation of different skin areas and a severe cellulitis of the subcutaneous tissue. The sudden and quick increase in death rates is often the first signal for the incidence of GD. The affected birds die after less than 24 h. The death rate is from 1% to 60%. The lesions are dark red to blue green macerated skin areas, usually featherless, beginning generally from wings and the adjacent areas. Most outbreaks are encountered in broilers at the age of 4-8 weeks. The disease is also observed in stock layer hens at the age of 6-20 weeks and broiler parents at the age of 20 weeks. The outbreaks are frequently observed in extremely wet and warm premises.

The gangrenous dermatitis affects birds while still alive. In more s evere cases, the gangrene could begin and involve the skin of the head, neck and the breast. Affected skin is macerated or totally necrotic, resulting in exposure of underlying tissues in a number of cases.

The GD agents are CI. septicum, CI. perfringens type A and Staphylococcus aureus, independently or in combination. The associated infection is more severe. The skin lesions are often crepitating and are detected in the regions of breast, abdomen, back or wings in both alive and dead birds.

Under the affected skin, extensive haemorrhagic oedemas with or without gas (emphysema) are discovered. The Clostridia are encountered in soil, faeces and the intestinal content. Staphylococci usually inhabit the skin and mucous coats of birds Underlying muscles are haemorrhagically infiltrated and also could contain gas among the muscle fibres. The increased susceptibility of birds to GD is related to immunodeficiency states secondary to CIA, IBD, IBH, etc. Other predisposing factors are aflatoxicoses, the unbalanced or deficient diet or poor hygiene. The skin wounds are the entrance door of the infection.

In most cases, no changes in viscera are observed. Rarely, gas bubbles could be found out in the liver. The diagnosis is confirmed by imprint preparations or histological cross sections. The bacterial agent could be isolated from cellulitic zones.

Single or multiple necroses are rarely seen in the liver. Prevention - the factors leading to skin injuries (cannibalism, mechanical traumas, etc.) should be minimised. Successful approaches are the immunoprophylaxis programmes in breeder flocks aiming at prevention and reduction of possible immuno¬suppressive states. Microscopic lesions are characterised by oedema, emphysema, hyperaemia, haemorrhages and necroses in the subcutaneous tissues. GD could be effectively treated by tetracyclines: doxyxycline hydrochloride, erythromycin thiocyanate, chlortetracycline and copper sulphate in the water.

Botulism

Botulism is an intoxication caused by the toxins of Clostridium botulinum. With the exceptions of vultures, most birds are susceptible. The clinical signs appear within a few hours to several days. Flaccid paresis of legs, wings, necks and eyelids is observed. The paresis is rapidly progressing to paralysis and the birds fall into a deep coma with neck and head typically extended forward. There are no typical gross lesions. The treatment with selenium, vitamins A, D and E as well as with some antibiotics as chlortetracycline, bacitracin, etc. could reduce mortality rate.

Spirochaetosis is a septicaemic disease characterised by depression, progressive paresis, paralysis and inflammatory necrobiotic changes in parenchymal organs and the gastro-intestinal tract. The birds are depressed, cyanotic; greenish diarrhoea with considerable amounts of urates is observed. The transmission of spirochaetosis is related to the distribution of the fowl tick Argas persicus that is the reservoir and main vector. Hens, turkeys, geese, ducks, pheasants, etc. from all ages are susceptible. At a later stage, paresis and paralysis are developing, the birds become somnolent and comatose.

After removal of plumage, the ticks, adhered to the skin, are discovered. The aetiological agent is Borrelia anserina, a highly motile spirochaete with 5-8 coils. The organism is not resistant outside the host and could exist only in some vector. A typical finding for spiro¬chaetosis is the enlarged spleen with marble-like appearance. A. persicus remains infective for over 430 days and transmit spirochaetes to its progeny. Spirochaetes could be identified in Giemsa-stained blood smears. In the late stage of the disease, spirochaetes could not be detected.

The liver is often enlarged, mottled with necroses of a various size and marginal infarctions. Usually, bile-coloured mucoid enteritis is observed. The arsenics as well as some antibiotics including tylosin, tetracycline and penicillin are effective for treatment of infected birds. The application of vaccines is successful, but the acquired immunity is short and revaccinations are needed.

Avian Tuberculosis

Avian tuberculosis is a chronic infectious disease characterized by the formation of granulomatous lesions in viscera, a progressive weight loss and death. It is usually encountered sporadically in birds reared in small yards, zoos and is a problem among caged exotic birds.

The aetiological agent is Mycobacterium avium, a very resilient and acid-resistant microorganism. It is resistant to temperature changes, drying, pH changes, to many disinfectants and survives in the soil for years. Along the small intestine, single or multiple sub-serously prominating tubercles are detected. In advanced cases, tubercles could be observed along the bone marrow of the femur or the tibia. The main mass of the tubercle is composed of epitheloid cells, and peripherally, there are foreign body-type multinucleated giant cells. In older granulomas, central coagulation or caseous necroses are present.

Characteristic features of avian tuberculosis are conglomerate tubercles (gross and histological appearance). The diagnosis is based upon the complex evaluation of history, persisting lethality in adult flocks and the pathoanatomical findings. Avian tuberculosis should be differentiated from neoplastic diseases, coligranu-loma (Hjarre's disease), pullorum disease, etc. The treatment is not advised, as the disease is contagious for men.

Viral Inclusion Body Hepatitis

The viral inclusion body hepatitis (IBH) is an adenovirus infection characterised by haemorrhages and dystrophic necrobiotic changes in the liver and kidneys, accompanied by intranuclear inclusion bodies. A characteristic macroscopic lesion is the enlarged, dystrophic liver with yellowish colour and crumbly texture.

IBH outbreaks are encountered primarily in meat type chickens, most commonly at the age of 3-8 weeks. IBH often occurs as a secondary infection to immunodeficiency resulting from other diseases (IBD, CIA). On the background of dystrophic liver changes, haemorrhages of various intensity and size are outlined, thus creating a variety of liver lesions. Throughout IBH outbreaks, several serotypes from the 12 known avian adenoviruses (AAVs) of group I are isolated. The sick chickens cany the virus in their excreta, kidneys, tracheal and nasal mucosa. The virus is resistant to many environmental factors and could be easily transmitted by a mechanical route. The transmission of adenoviruses is realised vertically by breeder eggs and horizontally, via excreta (mainly faeces). In a number of cases, the dominating lesion is the massive mottled or striated haemorrhages of the liver.

IBH is characterised by a sudden onset and a sharply increased death rate that reaches peak values by the 3rd - 4th day and returns back within the normal range by the 6th - 7th day. The total death rate is usually under 10% but sometimes could attain 30%. More rarely, macro-scopically visible necrotic foci could be detected in the liver.

The kidneys are enlarged, pale and mottled with multiple haemorrhages. Many instances of IBH, the amount of pericardial fluid is increased (hydropericardium).

Clinical signs could be observed only several hours prior to death occurrence. They consist in pale comb and wattles, depression and apathy. Sometimes, the skin is icteric. Often, ecchymoses and striated haemorrhages in skeletal muscles are observed.

Microscopically, extensive dystrophic changes and necroses of liver parenchyma are detected. In the nuclei of hepatocytes, basophilic or eosinophilic inclusion bodies are detected. The basophilic inclusion bodies are usually dense and occupy the entire nuclear inner space (arrow b), whereas the eosinophilic ones are round or irregularly shaped and surrounded by a light halo (arrow e). The diagnosis is based upon the typical gross lesions and the history records. A principal approach in IBH diagnostics is the histological investigation that helps to detect the intranuclear inclusion bodies. IBH should be distinguished mainly from IBD and chicken infectious anaemia (CIA). With regard to IBH prevention and control, the eggs of broiler parent flocks, where the disease is consecutively appearing in the progeny, should not be used for hatching. The access of wild birds should be prevented as they are potential carriers and distributors of the virus. The most important steps in IBH prevention are the control of IBD and CIA. There are neither vaccines, nor an effective treatment.

Haemorrhagic Enteritis of Turkeys

The haemorrhagic enteritis (HE) is an acute viral disease in young turkeys, characterised by a sudden onset, bloody faeces and various, but often high death rate. Blood discharge from the vent, fresh blood in faeces or melena could be observed.

HE is observed in turkeys at the age of 6-11 weeks but most commonly at the age of 7 - 9 weeks. As an exception, HE could be observed in turkeys under 4 weeks of age, presumably due to maternal antibodies. The small intestine, especially the duodenum, has a dark red colour and ramiform blood vessels prominating under the serous coat, and sometimes, haemorrhages are seen through the intestinal wall. HE is characterised by rapidly progressing clinical signs and disease course of about 10 days. The peak death rate is reached by the middle of the period. The entire death rate varies between 2-3% and 5-10% but could reach 60% as well. The turkeys are depressed, a sharp reduction in forage and water consumption is seen. The extent of lesions of the interstitial mucosa varies from hyperaemic to haemorrhagic, severe haemorrhagic or fibrinous necrotic inflammation.

The spleen of infected birds is typically enlarged, haemorrhagic, crumbly, motted or marble-like. The aetiological agent of HE is an adenovirus of group (type) II (HEV). The viruses or marble spleen disease in pheasants (MSDV) and avian adenvirus splenimegaly (AAS) that are serglogically similar, beling to the same group.

Later, the spleen reduces its size 2-3 times and acquires a specific silvery-grey colour. The serological studies show that HEV is widely distributed among adult turkeys. There is no evidence for vertical transmission. HEV is spread horizontally by ingestion of faeces of infected turkeys. Contaminated litter, equipment, shoes, other birds, rodents, etc. are also important in the transmission of the infection.

The liver is enlarged, crumbly and mottled with multiple haemorrhages, varying from petechiae to ecchymoses. Dead birds are in a good condition, with pale skin. Once having appeared in the farm, the infection is often recurring in other flocks. The HEV infection in turkeys results in a transient immunosuppression and consequently, a secondary E. coli septicaemia Less frequently, extensive necrotic foci, sometimes haemorrhagically infarcted, could be detected in the liver. The kidneys are enlarged and haemorrhagic. In many instances, the small intestinal and gastric serous coats are encompassed by petechial or striated haemorrhages. Hyperplasia of the white spleen pulp, necroses, subcapsular and periarteriolar haemorrhages are observed.

A characteristic diagnostic sign is the discovery of large acidophilic, rarely basophilic intranuclear inclusion bodies in the reticuloendothelial cells. The displaced and condensed nuclear chromatin pattern around the inclusion bodies often looks like a crescent. Similar inclusion bodies could be sometimes found out in lamina propria of the intestinal mucosa.

The staining of peripheral blood smears reveals a picture, characteristic for septicaemia. The typical gross lesions and the history of the disease allow making a tentative diagnosis for HE. The histological detection of intranuclear inclusion bodies in the reticuloendothelial cells of the spleen or intestines confirms the diagnosis. For identification of the HEV, the agar gel precipitation method could be used.

The viral antigen could be detected in fresh or frozen spleen tissue, diluted 1:1 with saline or in sera obtained from diseased turkeys. HE should be distinguished from other cases of severe enteritis in turkeys, acute E. coli septicaemia and some other septicaemic states (streptococcosis, fowl cholera, etc.). Within weeks after the beginning of the disease, an appropriate antibiotic therapy should be initiated for prevention of secondary E. coli septicaemia. Practically, there is no effective therapy for HE in turkeys.

Viral Diseases

Egg Drop Syndrome

The egg drop syndrome - 1976 (EDS 76) is an infectious disease in layer hens manifested by a quick drop in egg production, failure to reach peak production, irregularly shaped eggs, soft-shelled or shell-less eggs and depigmentation. The aetiological agent is an adenovirus of group III. The horizontal transmission occurs slowly in battery systems and rapidly in floor housing systems. The first sign is the loss of egg pigmentation, rapidly followed by the appearance of soft-shelled, shell-less of deformed-shell eggs. If defective eggs are discarded, the remaining ones have no problem with fertilization and hatching.

The drop could be sudden or prolonged. Usually, it lasts for 4-10 weeks and the egg production is reduced by about 40%. Apart the inactive ovaries and oviduct atrophy, other lesions are not discovered. The replication of the virus in epithelial cells of oviduct glands results in severe inflammatory and dystrophic changes in the mucous coat. The appearance of eggs with impaired quality and the dropped egg production are suggestive for EDS 76.

The diagnosis is supported by some serological studies and is confirmed after isolation and identification of the virus. In many instances, no antibodies are detected in infected flocks until egg production approaches levels between 50% and peak production.

Adenovirus Group I - Associated Infections

Necrotic pancreatitis in broilers. It is characterised by focal necroses in the pancreas. Gizzard erosions in broilers. They are manifested by erosions affecting the cuticle and the underlying tissues of the gizzard. Avian adenovirus splenomegaly. Avian adenovirus splenomegaly is characterised by enlargement of the spleen, pulmonary oedema and congestion. It is observed in broiler breeders at the age of 20-45 weeks. Its course is peracute or acute. The mortality could reach 8 - 9%. The most typical lesions are splenomegaly, mottled or marble-like appearance of the spleen, oedema or hyperaemia of lungs.

Infectious Bursal Disease (Gumboro)

Infectious bursal disease (IBD, Gumboro) is an acute, highly contagious viral infection in chickens manifested by inflammation and subsequent atrophy of the bursa of Fabricius, various degrees of nephroso-nephritis and immunosuppression. Clinically the disease is seen only in chickens older than 3 weeks. The feathers around the vent are usually stained with faeces containing plenty of urates.

The period of most apparent clinical symptoms and high death rate is at the age of 3 - 6 weeks. IBD could however be observed as long as chickens have a functioning bursa (up to the age of 16 weeks). In chickens younger than 3 weeks, IBD could be subclinical, but injured bursa leads to immunosuppression. Also, diarrhoea, anorexia, depression, ruffled feathers, especially in the region of the head and the neck are present.

A natural IBD infection is mostly observed in chickens. In turkeys and ducks it could occur subclinical^, without immunosuppression. Most isolates of the IBD virus in turkeys are serologically different from those in chickens. In premises, once contaminated with the IBD virus, the disease tends to recur, usually as subclinical infection. The dead bodies are dehydrated, often with haemorrhages in the pectoral, thigh and abdominal muscles.

The IBD virus belongs to the Birnaviridae family of RNA viruses. Two serotypes are known to exist, but only serotype 1 is pathogenic. The virus is highly resistant to most disinfectants and environmental onditions. In contaminated premises, it could persist for months and in water, forage and faeces for weeks. The incubation period is short and the first symptoms appear 2-3 days after infection. The lesions in the bursa of Fabricius are progressive. In the beginning, the bursa is enlarged, oedematous and covered with a gelatinous transudate.

IBD virus has a lymphocidic effect and the most severe injuries are in the lymph follicles of the bursa of Fabricius. Most commonly, IBD begins as a serous bursitis.

IBD lesions undergo various stages of serous haemorrhagic to severe haemorrhagic inflammation. The morbidity rate is very high and could reach 100%, whereas the mortality rate: 20 - 30%. The course of the disease is 5-7 days and the peak mortality occurs in the middle of this period. In some cases, the bursa is filled with coagulated fibrinous exudate that usually forms casts with the shape of mucosal folds. In birds surviving the acute stage of the disease, the bursa is progressively atrophying. Microscopically, an atrophy of follicles into the bursa is observed secondary to inflammatory and dystrophic necrobiotic alterations.

The kidneys are affected by a severe urate diathesis. In an acute outbreak and manifestation of the typical clinical signs, the diagnostics is not difficult. The diagnosis could be confirmed by detection of typical gross lesions throughout a patho-anatomical study. IBD should be differentiated from IBH (inclusion body hepatitis). The application of live vaccines in chickens is a key point in the prevention of IBD and should be related to the levels of maternal antibodies.

Infectious Bronchitis (IB)

In chickens up to the age of 4 weeks, IB is manifested by severe respiratory signs (sneesing, coughing, and rales). Rhinites and conjunctivites, depression and crowding around heat sources are observed. The death rate could reach 100%. The mortality in young chickens is usually insignificant provided that a secondary infection with a different agent is not occurring. In such cases, there is a moderate to severe inflammatory cell infiltration of upper respiratory tract mucosa, resulting In thickened and more compact mucosa.

In layer hens infected with the IB virus, oophorites and dystrophic necrobiotic lesions affecting primarily the middle and the last thirds of oviduct's mucous coat are observed. The consequences are drop in egg production, appearance and increase in the number of deformed and pigmentless eggs or eggs with soft shells and watery egg white.

The nephrotropic strains of the IB virus cause severe inflammatory and dystrophic necrobiotic damages of kidneys: urolithiasis, interstitial nephritis, haemorrhages (that considerably increase the death rate. Under natural conditions, only hens are infected. Non-immune birds of all ages are susceptible. The disease is even seen in vaccinated flocks. The serological methods (VN, ELISA, etc.) are widely used in the diagnostics. At present, PCR is used for rapid identification of IB virus serotypes. IB should be distinguished from other acute respiratory disease as ND, laryngotracheitis and infectious coryza. The vaccination with live or killed vaccines is effective only if they contain the respective serotype of the virus for the given region.

Laryngotracheitis

Laryngotracheitis (LT) is a viral infection in hens, pheasants and peacocks characterised by catarrhal haemorrhagic to fibrinous inflammation of the respiratory tract. It is manifested in laryngotracheal and conjunctival form. In the laryngotracheal form, suffocation, rales and cough are observed. The head and the neck are strongly extended forward and upward during inspiration. Mucous coats of the larynx and the trachea are catarrhally haemorrhagically to fibrinously inflamed. Most outbreaks are encountered between the age of 4 and 14 weeks although the disease affects fowl of any age. LT is caused by a herpesvirus that is relatively resistant. Haemorrhagic laryngotracheitis. The morbidity rate of LT reaches 50-70% and the death rate: 10-20%. Often, it goes on as a complicated infection after the involvement of E. coli, St. aureus, M. gallisepticum, etc. In some cases, casts of haemorrhagic or fibrinous exudate are formed that could almost completely obturate the larynx and the trachea. Source of the infection are sick and convalescent birds, the latter being prolonged carriers of the virus (up to 1 - 2 years). With this regard, a certain stationarity is observed.

In the conjunctival form of LT, wet eyes, tear secretion and oedema of infraorbital sinuses are observed, especially in a complicated infection. The typical clinical and morphological signs are sufficient to assume the presence of LT. The diagnosis is confirmed with the detection of intranuclear inclusion bodies in the trachea throughout the histological study in the early stages of the disease, serological studies (VN, ELISA), etc. LT should be differentiated from IB, SHS, M. synoviae infections, etc. Premises, contamined with the LT virus, should be freed, cleansed, disinfected and occupied again after 5 - 6 weeks. The vaccination of unaffected birds and these from other premises of the infected farm could protect and stop subsequent outbreaks.

Swollen Head Syndrome

Swollen head syndrome (SHS) is a complicated infection in broilers and broiler breeders, where the primary aetiological agent is an avian pneumovirus (APV) and the secondary -usually E. coli; it is characterised by respiratory and nervous signs. In broilers chickens, SHS is generally seen after the 4th week of life. The first clinical signs are sneesing, coughing, rales and conjunctivites. A profuse tear secretion, reddened conjunctivas and a characteristic oblong almondlike shape of eyes are observed. The inflammatory exudate is initially transparent, but afterwards becomes opaque. Subcutaneous oedemas in the head region, involving unilaterally or bilaterally the periorbital sinuses and the mandibular space, are emerging.

A seasonal pattern in the prevalence with peaks during spring and summer is observed. After removal of the covering skin, deposits of serofibrinous exidate are observed. APV is a virus from the Paramyxoviridae family. The pneumovirus is present in respiratory secretions and discharges. It survives for a long time in cold and wet environment. E. coli follows the APV infection of the upper respiratory tract. The triggered inflammatory response results in accumulation of exudate in the subcutaneous tissue. In many instances, a croupous pneumonia develops at a later stage consequently to contamination with other pathogens.

SHS in broiler breeders is usually encountered around or after the peak egg laying period only in female birds. Unilateral or bilateral swellings of the head, affecting the periorbital sinuses, the mandibular space and the wattles are seen. The conjunctiva and the mucous membranes of sinuses are considered to be the entrance door of the infection. A unique cytotoxin has been identified in many E. coli isolates in SHS that could be involved in the pathogenesis of the disease. The inflamed by the virus infection conjunctiva-associated lymphoid tissue is the site where the bacterium enters the subcutaneous tissue.

Frequently, nervous signs are observed in broiler breeders (opisthotonus, torticolis) due to inflammatory processes in pneumatic skull bones and the middle ear. In laying hens, the ovaries are affected in many instances (serofibrinous oophoritis), resulting in reduced egg production. The diagnosis is based upon the distinctive clinico-morphological signs. SHS should be differentiated from Mycoplasma and Pasteurella infections and the skin form of aspergillosis. Prevention - improvement of the microclimate of premises, use of live and killed vaccines.

Infectious Encephalomyelitis

The infectious encephalomyelitis (IEM) is characterised by signs of ataxia, progressing to paralysis, prostration and marked tremor of the head and the neck, and because of that, is also called epidemic tremor. The chickens with prostration are usually in lateral recumbency. The tremor could be unapparent, but is often perceptible when the chicken is held gently with the hand and carefully looked at. the expression of the eye is dull. IEM outbreaks are generally observed in chickens at the age between 8 - 9 and 20 days. the morbidity rate could reach 40 - 60%. the average death rate is about 25%, but could be more than 50%. Gross lesions are not present.

The histological lesions are specific and with a diagnostic value. A nonpurulent encephalomyelitis with marked perivascular clusters is present. The IEM virus is from the Picornaviridae family. The previous studies placed the virus in the Enterovirus genus, but based on the most recent studies, it is provisionally referred to the Hepatovirus genus. The virus is found in faeces of infected chickens and could survive there for at least 4 weeks. It is relatively highly resistant to environmental conditions.

A particularly valuable finding is the central chromatolysis of neurons in segments from the lumbosacral widening of the spinal cord and more rarely, chromatopyknosis. The infection occurs in non-vaccinated broiler breeder flocks and its course is subclinical. A vertical transmission of the infection to susceptible birds is realised by the eggs laid during this period. At the same time, parents acquire immunity and transmit it to the next generations thus protecting them from IEM. So, clinical manifestation of IEM is observed in chickens, hatched from the eggs during the 2- or 3 week period after breeder flock's infection. The next generations are immune.

Valuable diagnostic findings are the dense lymphoid clusters in the muscles of the proventriculus and the gizzard and in the pancreatic interstitium. The history of the disease, the age of onset and the typical nervous signs, especially the head tremor are indicative for the diagnosis that could be finally confirmed by histology. IEM should be differentiated from other diseases with nervous symptoms as encephalomalacia, mycotic encephalites, toxicoses (salt, pesticides). Prevention -vaccination of breeder flocks with regard to ensuring maximum protection of their offspring. The chickens from naturally infected flocks obtain an adequate maternal immunity that preserves them from the disease.

Newcastle Disease

The Newcastle disease (ND) is a highly contagious disease in many species of domestic, exotic and wild birds that, depending on its tropism, is characterised by marked variations in morbidity, death rate, symptoms and lesions. The clinico-morphological signs possess a distinct viscerotropic or neurotropic character. In the viscerotropic form, haemorrhagic diphtheritic lesions of the entire alimentary tract, from the beak to the vent, are present. The haemorrhages of gizzard epithelium are remarkable. The mucous coat is oedematous, covered with thick mucus and mottled with haemorrhages varying from single to multiple, sometimes gathered at the boundaries with the gizzard or the oesophagus.

Typical for this form are the haemorrhagic necrotic and focal diphtheroid lesions affecting the mucosa of the buccal cavity, the stomach and the intestines.

The disease is generally prevalent in hens, more rarely in turkeys, exotic or wild birds. Birds at any age are susceptible. It is caused by a paramyxovirus. Depending on their pathogenicity, the numerous known strains are classified as lentogenic, meso-genic and velogenic. The vaccines made of lentogenic strains provoke a shorter immunity that requires a revaccination. The vaccines from mesogenic strains result in a lasting immunity, but could provoke a lethal issue especially in birds without a primary immunity created on the basis of lentogenic vaccinal strains.

A frequent finding is the enlargement and haemorrhages of caecal tonsils and haemorrhagic cloacitis. Usually, these lesions begin from the lymphoid tissue of the mucous coat. Virus-containing excreta of infected birds, that contaminate the forage, water and the environment, are the source of infection.

The infection is transmitted mainly by an oral route, the airborne or contact transmission being more infrequent. The virus, contained in incubated eggs, results in embryo's death and then perishes. There is no permanent carriership of the virus. An important factor in the transmission of velogenic viruses could be the exotic birds and fighting cocks. The death rate could arrive at 70-100%.

The neurotropic form of the disease is clinically manifested by ataxia, opisthotonus, torticolis, paresis and paralysis of legs. This form is frequently accompanied by respiratory symptoms. Histopathologically, the picture of non-purulent lymphocytic encephalomyelitis is observed. The lesions of paramyxovirosis in pigeons are entirely identical. On the basis of the history and the clinicomorphological signs, a tentative diagnosis could be made, but its laboratory confirmation is mandatory. ND should be distinguished by avian influenza, fowl cholera, etc.

Fowl Pox

Fowl pox (FP) is a viral disease in hens, turkeys and many other birds, characterised by cutaneous lesions on the feather-less skin and/ or diphtheritic lesions of mucous coats of the upper alimentary and respiratory tract. FP is encountered in either cutaneous or diphtheritic form or in both. In most outbreaks, the cutaneous form is prevailing. The lesions vary according to the stage of development: papules, vesicles, pustules or crusts. The lesions are usually in the region of the head. Lesions around the vent in a pigeon. FP is caused by an epitheliotropic DNA virus from the Avipox genus, the Poxviridae family. Some virus types (strains) exist: fowl pox virus, turkey pox virus, pigeon pox virus, canary pox virus, etc., different in pathogenic and immunogenic aspects. The viruses are very resistant to environmental factors and could persist for several months.

Frequently, the conjunctival mucosa, injured by the pox virus, is an entrance door for additional contamination (£. coli, Staphylococcus spp., etc.) and development of complications. The infection is mechanically spread by dissemination of the virus through desquamation of crusts that contain it. Some mosquitoes and blood suckling arthropods could also distribute the virus. The mosquitoes remain infective for several weeks. The incubation period is from 4 to 10 days. The disease is spread slowly and many weeks could pass between its emergence and severe outbreaks occurrence.

Diphtheritic lesions look like whitish or yellowish plaques that are deposited and grown on the mucous coats of the buccal and nasal cavities, the sinuses, the larynx, the pharynx, the trachea or the oesophagus (arrows). The diagnosis is made upon the typical cutaneous and diphtheritic lesions. The prevention is performed by vaccinations that could be made at any age, if necessary.

Reovirus Infections

Viral Arthritis/tenosynovitis: The most frequent reovirus-associated disease in poultry is viral arthritis. Clinically it is manifested by lameness and swellings affecting primarily tarsometatarsal joints and the feet. Many infected birds are in a good general condition, but some could be lethargic and exhausted. The mortality is very low. The infections affect predominantly meat type poultry.

In some cases, joint cavities or tendon sheaths contain a small amount of straw-yellow exudate whereas in other the exudate is haemorrhagic or fibrinous.

Fibrinous Tenosynovitis: The inflammation of the tendon progresses to a chronic type lesion characterised by tissue fibrosis in the affected area. The aetiological agent is a reovirus, member of the Orthoreovirus genus, Reoviridae family. Several serotypes are identified. For their identification, the agar gel precipitation test could be used. Reoviruses are highly resistant to a number of environmental factors (temperature, pH, etc.).

Stunting Syndrome in Broilers: The stunting syndrome in broilers is associated with a reovirus infection but according to some studies, the role of the reovirus is probably secondary. It is characterised by a considerably reduced live weight in affected birds and a various degree of ununiformity in the flock varying from 5-10% to 40-50%, usually seen after the age of 14 days.

The growing primary wing feathers are abnormally big for chickens with retarded growth, they protrude at various angles, so the disease is termed helicopter disease". One-day old chickens are the most susceptible to the infection.

Usually, a High-degree Atrophy of the Pancreas is Observed: Reoviruses are shed with faeces and could contaminate the egg shells. The transmission of the infection to susceptible chickens is realised horizontally. The vertical route of transmission is also proved. Reoviruses could persist in infected birds for more than 40 weeks.

The Small Intestine is Pale, Dilated and Contain Indigested Forage: The tentative diagnosis of viral arthritis and stunting syndrome could be made on the basis of symptoms and lesions. The detection of reovirus antibodies via ELISA supports the diagnosis. Viral arthritis should be differentiated from M. synov/ae-induced, staphylococcal arthrites and the spontaneous rupture of the tendon of the gastrocnemius muscle. The vaccination of broiler breeder flocks with live or inactivated vaccines protects one-day-old chicks. Infected birds should be removed from the premises. The iodine disinfectants are considered effective for inactivation of agents.

Neoplastic Diseases in Poultry

Virus-induced Neoplastic Diseases Marek's Disease

Acute (visceral) form. It is characterised by diffuse or nodular lymphomatous lesions in various viscera (liver, spleen, heart, kidneys, lungs, gonads, proventriculus, pancreas, etc.), the skeletal muscles and the skin. MD affects mainly hens, and is rarely observed in turkeys. It is most commonly encountered in birds at the age of 89 weeks and in layer hens. The cases at the age of 16-20 and 24-30 weeks are pre¬dominant. MD is prevalent all around the world and in fact, all flocks are exposed to the effect of the aetiological factor.

Diffuse lymphomatous growths in the heart, resulting in its transformation into an amorphous tumourous mass. Bilateral enlargement of kidneys because of a diffuse lymphoid cell proliferation.

Neoplastically modified right lung in MD: Typical cauliflower-like appearance of the ovary, distinctive for MD. Marked asymmetry of testes on a cock following unilateral lymphoid cell proliferation.

Diffuse neoplastic growths affecting the pancreas: The manifold enlarged proventriculus with the shape of a round bottom flask result of diffuse neoplastic growth and the severely narrowed lumen are a typical finding in MD. The causative agent of MD is a type B cell associated herpesvirus (MDV). There are three MDV serotypes. The isolates of serotype 1 are widely distributed among hens and vary from highly virulent (w+) oncogenic to almost avirulent strains. The serotype 2 is common for hens and is not oncogenic. The isolates of serotype 3, known also as turkey herpesviruses (HVT) are naturally occurring in turkeys and are non-oncogenic. The three serotypes possess a significant cross reactivity.

Multicentric MD tumours (arrows) prominating or seen through the superficial and deep pectoral muscles.

Chronic (classical) form: It is ecountered as neural type (fowl paralysis) or ocular type (ocular lymphomatosis). Clinically, the neural for is manifested with paralysis of limbs.

Pathoanatomically, unilateral or bilateral thickening of affected nerves, mainly diffuse and at a various extent, is observed. The ocular form is characterised with iris depigmentation, deformation of the pupil, sometimes opacity of the cornea and blindness. Histologically, pleiomorphic lymphoid cell proliferations in affected viscera, nerves or eyes are observed.

The Marek's disease is probable provided that at least one of the next conditions is present: peripheral nerves augmentation, depigmentation of the iris or irregularly-shaped pupil; lymphoid tumours in various organs in birds younger than 16 weeks; presence of visceral tumours in birds at the age of 16 weeks and older; simultaneous lack of alterations in the bursa of Fabricius. Microscopical view of lesions in a peripheral nerve consequent to MD.

Lymphoid cell proliferations in the iris and the ciliary muscles in the ocular form of MD. Three classes of viruses are able to protect fowl from MD: attenuated serotype 1 of MDV-cell associated vaccines; HVT could be used for preparation of cell-free lyophilised vaccines; naturally apatho-genic isolates of serotype 2 cell associated vaccines. The vaccines against MD provide over 90% protection. HVT gives excellent results but in case of failure, a bivalent vaccine could be used.

Transient Paralysis: They are observed in chickens and hens, especially non-vaccinated against MD. Most cases present the classical form manifested by flaccid paralysis of the neck and legs for 143 days followed by complete recovery. The syndrome has to be differentiated from the neural form of MD on the basis of its transient nature and the flaccid, but not spastic paralysis.

Lymphoid Leukosis: It is characterised by a gradual beginning, persistent low mortality in the flock and diffuse or focal neoplastic growths of lymphoblasts in viscera. The neoplastic changes begin always from the bursa of Fabricius, where various-sized lymphomas are detected (transverse section through neo-plastically grown bursa -fixed preparation).

Clinically, pale comb and wattles, sometimes swelling of the abdomen because of the highly enlarged liver are observed. Diffuse or nodular neoplastic growths could be detected in many organs, but they are more common in the liver, the spleen, the kidneys, the heart and the ovary.

Spontaneous rupture of the neoplastically grown spleen, leading to extensive loss of blood. LL is widely distributed worldwide in countries with developed industrial poultry breeding.

It is usually observed in birds at the age of 16 weeks and older. Focal neoplastic lesions in kidneys. LL is caused by viruses of the L/S group classified in 10 subgroups: A, B, C, D, E, F, G, H, I and J. The viruses from subgroup A are most prevalent and most frequently associated with LL. Hens, rarely turkeys, pheasants and quails are susceptible.

Diffuse and focal tumour lesions in the heart. The replication of the virus occurs in albumin secreting glands of the oviduct. The transmission of the infection is performed vertically by egg albumin from one generation to another. The role of cocks is not important for the congenital infection of the progeny. They are only virus carriers and source of venereal infection for other birds.

Neoplastically transformed ovary in LL. In some instances, the horizontal infection is also possible but only in chickens in the first few days after hatching, usually via vaccines contaminated with ALSV. The lethal issues are observed for 56 months after the LL outbreak and amount to 5 -15%. Histologically, growth of single type lymphoblast cells with marked pyroninophilia is observed. In the bursa of Fabricius, a characteristic intrafollicular hyper¬plasia is observed.

The picture of an imprint preparation from neoplastic lesions shows a layer of single-type lymphoblast cells. LL and MD are hard to be distinguished: in both, lymphoid tumours are present in the same visceral organs, the appearance at the same age is possible, and the visceral lesions could not be differentiated macro-scopically, except in a careful microscopic examination by an experienced pathologist.

From the Point of View of Differential Diagnosis, the Following Features Deserve a Special Emphasis

Lymphoid Leukosis

- Usually, it is not seen in birds younger than 14 weeks.
- The lethal issues occur mostly at the age between 24 and 40 weeks.
- Distinct nodular tumours.
- Tumours in the bursa of Fabricius.

Marek's Disease

- Could be observed after the age of 4 weeks too.
- The peak mortality is seen between the 10th and the 20th week, sometimes continues after the 20th week.
- Paralysis.
- "Grey eye".
- In some birds, the bursa of Fabricius is atrophied, in others: neoplas

Myelocytomatosis

MC is caused by viral staings of ALSVs from sungroups A,B and J (Mc29, MC31, Cm11, OK10, HRPS 103, and ADOL HC1). It is encountered relatively infrequently. Its occurrence is sporadic or enzootic. Suceptible birds and hens, pheasents, guinea hens and quails. In most cases, the liver is enlarged, thick and mottled with dark red sports or fat like nodules.

MC is caused by viral strains of ALSVs from subgroups A, B and J (MC29, MC31, CMII, OK10, HRPS 103, and ADOL HC1). It is encountered relatively infrequently. Its occurrence is sporadic or enzootic. Susceptible birds are hens, pheasants, guinea hens and quails. In most cases, the liver is enlarged, thick and mottled with dark red spots or fat-like nodules.

Sclerotic changes in the liver are possible because of regression of neoplastic lesions. The spleen is usually enlarged, but sometimes, could be atrophied.

A characteristic feature of MC is its simultaneous course with tumours from a different type: mesenchymal, epithelial or mixed. The picture shows a fibrosarcoma to the gizzard associated with MC. Mixed mesenchymal tumour (osteochondrosarcoma) to the frontal skull bones: a sagittal cross section

Multiple rabdomyosarcoma in pectoral, thigh, abdominal and tracheal muscles.

Leiomyosarcoma of the mucous coat on the oviduct.

Pendulation harmangio-sarcoma of the ileal serosa.

Pendulating multiple myxoma of the small intestine's serous coat.

MC-associated cystadenocarcinoma of the kidney in a hen.

Nephroblastoma of the left kidney, occupying a significant part of the abdominal cavity. Nephroblastoma -the surface of a cross section. The tumour is a pendulating mass attached to the kidney by a fibrous vascularised stem that has undergone a partial necrosis and haemorrhages. Granulosa cell tumour of the ovary. The tumour appears as a single, compact, dorsoventrally flattened growth.

MC-associated multiple carcinosarcoma of the mesentery and alimentary tract's serous coat (disseminated milliary nodules).

MC-associated carcinosarcomas in the region of the right infraorbital sinus.

Histologically, myelocytomatomas are easily distinguished. Most commonly, they have perivascular localisation. Growth of myelocytes with well-formed granules in a liver cross-section.

Kidney Focal intertubular myelocytic proliferations.

MC-associated neoplasms of epithelial, mesenchymal or mixed type demonstrate the respective type of histological structure. Leiomyosarcoma a histological view. Polygonal giant cells with hyperchromatic nuclei. Leiomyosarcoma - small intestime. Prolongations of polynuclear symplastic elements. Leiomyosarcoma - small intestine. Extraordinary („monstrous") multinuclear giant cell with intracytoplasmic vacuoles. Rabdomyosarcoma. An area with multiple hyperchromatic giant cells.

Carcinosarcoma of the pancreas. Tubulous glandular epithelial formations of the carcinoma component among the liposarcoma part of the parenchyma. The diagnosis is based upon the entity of data about the history, the gross appearance and location of the tumours and the specific histological lesions. From a differential diagnostic point of view, myeloblastosis and erythroblastosis should be considered.

Erythroblastosis

Erythroblastosis (ER) is characterised by intravascular proliferations of immature precursors of erythrocytes. ER has a leukaemic character and is manifested with signs of severe anaemia. The liver and the kidneys are moderately enlarged with a characteristic dark red to mahogany colour, sometimes with haemorrhages. ER is caused by the avian eryhtroblastosis virus (AEV); the most frequently encountered strains are E-26, ES4, R, etc. The spleen is unusually enlarged or atrophied in cases of severe anaemia.

Histologically, accumulation of erythroblasts in blood sinusoids and capillaries is seen. The diagnosis is based on visceral histological lesions, typical for ER and peripheral blood haematological and morphological analysis. Osteopetrosis is a neoplastic disease, aetiologically related to the L/S group of viruses. It is characterised by a significant thickening of bone periosteum. The diaphyses of the tibia and/or tarsometatarsal bones are most commonly affected. Often, osteopetrosis is seen simultaneously with LL in the same bird.

Adenocarcinomatosis

The intestinal and reproductive tracts are primarily affected. Quite often, the neoplasms invade the peritoneum and other serous coats. Macroscopically they appear as numerous disseminated thick nodes of a various size (with a diameter from several mm to 1 cm).

Sometimes, the tumours are cystic formations (cystadeno-carcinoma). The leiomyoma of the mesosalpinx is a common tumour in hens. Usually, it is located in the ventral ligament of the oviduct. Its size varies from small to large (several cm in diameter), smooth, thick, sometimes highly vascularised nodes.

Parasitic Diseases

Coccidiosis

Coccidiosis is a common protozoan disease in domestic birds and other fowl, characterised by enteritis and bloody diarrhoea. The intestinal tract is affected, with the exception of the renal coccidiosis in geese. Clinically, bloody faeces, ruffled feathers, anaemia, reduced head size and somnolence are observed. The area around the vent is stained with blood.

The infection is realised by a faecal-oral route. After ingestion of sporulated (infective) oocysts, sporozoites are released that enter asexual and sexual cycles of development resulting in the emergence of thousands of new oocysts in the intestines. Oocysts are distributed by faeces. Soon, they sporulate and become infective for chickens.

The intestinal lesions provoked by coccidia, are due to injury of the epithelial cells of the mucous coat where the parasites are developed and multiplied.

The oocysts exist in the litter in premises and are distributed by clothes, shoes, dust, insects, etc. Pathoanatomically, dehydration and a high degree of anaemia of the body and viscera are discovered. Anaemic appearance of internal organs. The wet litter and the heat in premises favour of the sporulation and therefore, the outbreak of coccidiosis.

Depending on the localisation of lesions in intestines, the coccidioses are divided into caecal, induced by E. tenella, and small intestinal, induced by E. acervulina, E. brunetti, E. maxima, E. mitis, E. mivati, E. necatrix, E. praecox and E. nagani. In caecal coccidiosis, a marked typhlitis is present and haemorrhages are seen through the intestinal wall.

The caeca are filled with fresh or clotted blood. A later stage, the caecal content becomes thicker, mixed with fibrinous exudate and acquires a cheese like appearance.

In small intestinal coccidioses, depending on the eimeria species, haemorrhages with various intensities in different parts along the intestine are observed. In many instances, the haemorrhages are petchial and could be seen through the intestinal wall. Sometimes, a reation of the intestinal lymphoid tissue is present.

The content is mixed with fresh or clotted blood, and the mucous coat is mottled with multiple petechial or larger haemorrhages.

Histologically, Eimeria organisms at a various stage of development are detected in the epithelial intestinal cells. The diagnosis is made upon the results of the complex evaluation of the clinical picture, the macroscopic lesions, imprint preparations, histological study and flotation. Coccidioses should be differentiated from NE, UE and histomonosis (typhlohepatitis). Treatment - sulphonamides are widely used: sulphadimethoxine, sulphaquinoxaline, I sulphamethazine, but they should not I be used in layer hens. The supplementation I of vitamins A and K promotes the recovery.

The microscopic examination of a native preparation of intestinal content or superficial mucosal layer reveals a significant number of oocysts in one observation field. Prevention. The use of coccidiostatics with forages on a rotation basis is the most extensively used means. The immunisation against coccidiosis with commercial vaccines is used in broiler breeder flocks. If the chickens are exposed to the natural effect of a moderate number of oocysts in their environment, they develop immunity to the respective parasitic species.

Histomonosis

The histomonosis is a protozoan disease, caused by Histomonas meleagridis, and characterised by necrotising lesions affecting the liver and the caeca. Clinically, sulphur-yellow coloured faeces and depression are observed. A characteristic feature is the blackening of the skin of the head (blackhead), due to cyanosis.

Pathoanatomically, bilateral enlargement of caeca with thickening of walls is observed. The aetiological agent is Histomonas meleagridis, a polymorphic flagellate that is present as flagellate in caeca and as amoeba in tissues. The trophozoites survive for several hours in the environment but in Heterakis eggs, they remain infective for more than a year. Often, the occurring typhlitis is the cause for adhesive peritonitis. Susceptible species are turkeys, chickens, pheasants, rock partridges, guinea fowl, and geese. The turkeys are the most vulnerable between 3 and 12 weeks of age and chickens between 4 and 6 weeks of age. The caecal mucosa is usually ulcerated.

The main vector is Heterakis gallinarum through the eggs, respectively the larvae, where Histomonas meleagridis forms are found. Some wild birds could also serve as vectors. The caecal content is often mixed with blood. In older cases, crusts of dense caseous masses are formed into the carca that thicken this intestinal wall and reduce the lumen (top right: transverse cross section through caeca.)

In older cases, crusts of dense caseous masses are formed into the carca that thicken this intestinal wall and reduce the lumen (top right: transverse cross section through caeca.) Earth worms are mechanical vectors of H. gallinarum larvae. The main reservoirs of infection are hens and chickens. The morbidity rate amounts to 90% and the mortality rate to 70%. In the liver, irregularly outlined coagulation necroses with various size and colour, are observed.

Usually, necroses represent yellowish to grey or red (haemorrhagically infarcted), well delineated oci with diameter of about 1-2 cm. Diagnosis - it is made on the basis of the typical macroscopic lesions. When necessary, a histological study and phase-contrast microscopy of native preparations could be performed. Histomonosis should be differentiated from UE, coccidiosis and alimentary tract trichomonosis (Trichomonas gallinarum), where not counting the caeca, lesions are also present in the last third of small intestine.

Ascaridiosis

Ascaridiosis is one of the most prevalent helminthoses in fowl. It is caused by various species from the Ascaridia genus. The ascarids have a direct life cycle. Sometimes, it could involve paratenic hosts (earth worms). Infected birds are progressively emaciated, anaemic and sometimes diarrhoeic.

Pathoanatomically, haemorrhages of various intensities are found out in intestinal mucosa, catarrhal haemorrhagic enteritis and the parasites themselves are also observed. In cases of extensive invasion, the ascarids block the intestinal lumen and could cause a complete obstruction. The treatment and the control are realised through regular pathoana-tomical and coproovoscopic studies and performance of therapeutic and protective dehelminthisations.

Kraillietinosis

Raillietinosis is a cestodosis characterised by diarrhoea (sometimes bloody) during the acute stage and emaciation to cachexia and anaemic during the chronic stage. It is caused by some representatives of the Raillietina genus that parasitise in various areas of the small intestine. The usual intermediate hosts are ants or other insects. Pathoanatomically, haemorrhages with various intensities in the intestinal mucosa, catarrhal haemorrhagic enteritis and the parasites themselves are found out throughout the gross examination. The treatment and the control are done by dehelminthisation of all birds in the affected farm.

Knemidokoptosis

Knemidokoptosis, known also as scaly or chalky legs, is characterised by the appearance of extensive, rough and hard crusts on the featherless part of the legs. Adult birds are generally affected (hens, turkeys, pheasants, exotic birds). It is caused by Knemidokoptes mutans. The lesions are secondary to an inflammatory reaction, in which the thickening of the cutaneous epidermis together with released exudate forms the crusts on legs. The infection occurs by contact between birds. The control is done by isolation of affected birds, leg bathing with warm acaricide solution and warm vegetable oil that helps the decrustation.

Mycoses and Mycotoxicoses

Aspergillosis

Aspergillosis is an acute or chronic respiratory disease. In rare instances, peritoneal, visceral or systemic lesions could be observed. It is caused by Aspergillus fumigatus. Dyspnea, enhanced, tense and heavy breathing are observed. Sometimes, rales and cyanosis could occur.

In the acute form, a serous fibrinous pneumonia is observed. Within the trachea and the main bronchi, usually near the bifurcation, obturation masses of coagulated fibrinous exudate are detected. In the nodular form, multiple grey whitish or yellowish dense nodes in the lungs are observed. Chickens, turkeys, waterfowl and many other domestic or wild fowl are susceptible. Aspergillus granulomas of the serous coats: pectoral, abdominal air sacs and pleural, are sometimes compressed. The infection could occur inside (incubatory aspergillosis) or outside the hatchery. Aspergillus fumigatus penetrates through the egg shell under ideal conditions for development and infects the embryos.

The eggs explode and release a significant number of spores that contaminate the hatchery and the environment. The spores are spread via the ventilation system that results in severe outbreaks in chickens younger than 3 weeks. Outside the hatchery, the infection occurs by inhalation of numerous spores from the contaminated forage, litter or environment. The overcrowding and the increased humidity are prerequisites for outbreaks.

Rarely, as a systemic manifestation, Aspergillus granulomas in the brain could be detected, when spores are transported via the blood circulation. The signs and macroscopic lesions are very indicative for aspergillosis. They could be confirmed by histology too. In acute aspergillosis, among the inflammatory necrotic masses, the spores are observed (arrow a) as well as the grown hyphae (arrow b) of the mould.

In the nodular form, a characteristic Aspergillus granulomatous structure of tissues is discovered. The control includes: collection of clean eggs for hatching, disinfection and fumigation of eggs, incubators and hatcheries, regular checking of ventilation systems and change of air filtres in hatcheries, monitoring of hatcheries and the environment for mould contamination, use of dry and clean litter and non-contaminated forages, optimisation of aeration and the humidity in premises where birds are housed.

The treatment with fungicide antibiotics and different antimycotic drugs is not always efficient. Nistatin and copper sulphate (1:2000) with drinking water could be used. Mould is developed after inoculation of material on an appropriate nutrient media.

Aspergillus Granulomatous Dermatitis

The Aspergillus granulomatous dermatitis as a post-vaccinal compli¬cation is observed in growing broiler breeders. Following subcutaneous application of oil adjuvant vaccine in the lower part of the neck, a marked local reaction with extensive swelling of the head is observed.

The swellings involve the subcutaneous tissue of periorbital sinuses, the mandibular space, even the entire skull without the skin adnexa. The head is ballooned with strongly stretched skin. In some cases, the covering skin acquires a blue-greenish colour. The eyeball is sometimes affected by necrotic colliquation. Among the cross sectional surface if the highly oedematous subcutaneous tissue, multiple structures resembling sand grains are detected.

A later stage, a spontaneous regression of oedemas occurs, but granulomatous formations continue to be visible in the subcutaneous tissue. Histologically, among the oedematous subcutaneous connective tissue, multiple granulomas are detected. Central zone of granulomas is an inflammatory necrotic detritus with many eosinophilic leukocytes delineated within.

As a crown, multinuclear giant cells of a foreign-body type are arranged, with fibroblast fibrocyte growths around them. At a later stage of the condition, after phagocytosis of the detritus, the structure of granulomas is mainly composed by foreign-body giant cells and connective tissue growths.

Candidiasis

Candidiasis is an alimentary tract disease, caused by the fungal yeast Candida albicans and usually occurs as a secondary infection. Predisposing factors are the lack of good hygiene, prolonged antibiotic treatment, vitamin deficiency, severe parasitic infections and immunodeficiencies. The production of a soluble endotoxin is also contributing to the pathogenicity of the agent. The lesions are usually detected in the crop, lips, oesophagus, but could affect the proventriculus and more rarely, the intestines. The affected mucosa is diffusely or focally thickened, raised and wrinkled, white, resembling a towel. Possibly, the lesions could look as pseudomembranous or diphtheroid coatings. The histological study confirms the diagnosis by showing the fungal hyphae in the affected mucosa. Usually, a 1:2000 copper sulphate solution is used with drinking water for prevention and control. The application of nistatin in water or forage is efficient against candidiasis in turkeys.

Fusariotoxicoses

The Fusarium genus produces numerous mycotoxins, out of which, the most important for poultry pathology are trichothecenes, fumosinins, moniliformin, fusaro-chromanone and zearalenone. The Fusarium toxins possess a pronounced caustic effect, resulting in necroses and crusts of the buccal mucosa. The caustic effect of some fusariotoxins is the cause of the commonly detected erosions and ulcers in gizzard cuticulum (arrow a). Note the thickened wall of the proventriculus (arrow b). The cuticulum of the gizzard in broilers whose diet contains Fusarium toxins, is frequently with a linear appearance and under the ulcerations, the subcuticular surface is seen.

The high degree of spleen atrophy (arrow) is a common finding, a sign of the immunosuppressive effect of mycotoxins. The hyperaemic and haemorrhagic mucous coat of the gizzard is a frequent finding in fusariotoxicoses. The lesions include also reddening and hemorrhage of intestinal muscosa. The size of the hemorrhage varies from 23 to 56 mm in diameter and they could often be seen through the intestinal wall. Usually, the mucous coat of the duodenum and the initial part of the ileum are affected.

A significant part of dead bodies are dehydrated. Inside the liver, hemorrhages with various intensities outlined at the background of severe dystrophic changes, are detected. Frequent findings in fusariotoxicoses are the massive subcapsular liver haematomas, causing sudden death in broilers. The fusariotoxin zearalenone has an effect, identical to that of oestrogenic hormones and results in reduction of testes in cocks. Left - normal; right - atrophied testis in a cock, in whose diet high zearalenone concentrations have been determined.

Microscopically, the testes of cocks with zearalenone fusariotoxicosis, show a fatty infiltration and atrophy of the germinative epithelium with the exception of the basal layer as well as interruption of the spermatogenesis.

Fusarochromanone causes tibial dyschondroplasia in broiler chickens, manifested with long bone deformation.

Frequently, the proventriculus is dilated at a various extent, its wall is thickened and in its mucous coat, fibrinous necrotic and haemorrhagic lesions are detected. The prevention from mycotoxicoses requires the detection and control of mycotoxin-contaminated forage components, avoidance of forage moulding and thus, the formation of mycotoxins.

A screening of cereals and forages for the presence of some mycotoxins (aflatoxin, T-2 toxin, zearalenone) via ELISA is advised. The application of commercial mycotoxin-binding agents could possibly reduce the effects of some of them. In some mycotoxicoses, the systemic requirements for vitamins, minerals and proteins are increased and they could be balanced by supplementation with forage or water.

Deficiency Diseases

Vitamin Bi Deficiency

In birds, vitamin B, (thiamine) deficiency is clinically and morphologically manifested with paralysis of limbs and muscle atrophy, beginning from the flexors of toes and ascending towards the extensors of legs, wings and the neck. The chickens acquire a specific posture with flexed legs and the head drawn back (stargasing). As hypovitaminosis B, causes a severe anorexia, thiamine supplementation to the drinking water is advised until the recovery of acute deficiency and thereafter, the vitamin could be added to the forage.

Vitamin E Deficiency

The deficiency of vitamin E in poultry is manifested in three different forms: encephalomalacia, muscular dystrophy and exudative diathesis. Each of them is usually encountered independently, although sometimes they occur at a time. Encephalomalacia (crazy chick disease).

The signs are associated to the central nervous system lesions. They include ataxia, disequilibrium, falling on the back, frequent movements with the wings, strongly stretched legs (clonic spasms) and twisting of the head. Rarely, torticolis or opisthotonus could be observed.

Encephalomalacia is usually observed at the age of 15-30 days but could be also present as early as after the 7th day, as well as after the 56th day of life.

Pathoanatomically, oedema, haemorrhages and colliquative necroses are detected in the cerebellum. In the majority of cases, the haemorrhages vary from hardly perceptible to petechial.

It is possible, although more rarely, to observe massive haemorrhages and sometimes, haematomas in the cerebellum. As an exception, brain lesions could also be present. The vitamin E deficiency is usually manifested in young birds -chickens, turkey poults, ducklings, pheasant poults, etc. Most outbreaks are related to high levels of polyunsaturated fat in the diet (meat and bone meal, fish meal, etc.) or rancid fat content.

Histologically, colliquative necroses in the white brain substance, haemorrhages and multiple thrombosed blood vessels in the cerebellum are detected. Vitamin E and the selenium-containing enzyme glutathione peroxidase preserve the cellular membrane from being destructed by peroxides and other oxidants, produced as metabolic by-products. Peroxides are derivatives of polyunsaturated fatty acids in forages.

Muscular dystrophy. The signs are usually unapparent, but locomotor problems could occur. The pectoral and thigh muscles are most commonly affected, and in them, white or yellowish muscle fibres are found, responsible for the striated appearance of skeletal muscles, due to Zenker's necrosis.

Exudative diathesis. Gelatinous subcutaneous oedemas are discovered in the ventral body parts: breast, abdomen, mandibular space. The skin of legs is often cyanotic. The diagnosis is based upon the typical clinical signs and the lesions.

The results of the histological examination have a particular value for confirmation of the diagnosis, especially of encephalomalacia and muscular dystrophy. Control -utilisation of stabilised fats in forages, avoiding the prolonged storage of ready forages for more than 4 weeks. The vitamin E content should be 10 000 lU/tonne and that of selenium - 0.25 ppm.

Fatty Liver Haemorrhagic Syndrome

The fatty liver haemorrhagic syndrome (FLHS) is a widely prevalent sporadic disease mainly among commercial layers. The FLHS outbreaks are often associated with hot weather and a period of extensive egg-laying. The hens in the flock are overweight (on the average by 20% or more) and a sudden drop in egg production is observed. The birds are discovered suddenly dead, with pale head skin. In the abdomen, large blood clots are detected.

The liver is enlarged, pale and brittle.

In other instances, the organ is yellow, greasy and soft.

Subcapsular parenchymal haematomas are possible. It is assumed that high energy forages and the restricted locomotion are prerequisites for fattening of the liver. Other possible contributing factors are the deficiency of lipotropic agents, necessary for fat mobilisation by the liver, aflatoxins, genetic factors, etc. Frequently, FLHS and cage layer fatigue are diagnosed at a time.

Clinically healthy birds in the flock could also exhibit liver haematomas, dark red (fresh) or green to brown (old). Considerable amounts of fat are detected in the abdominal cavity. The only successful approach for prevention is the reduction of obesity in layers. The use of lipotropic agents such as vitamin E, vitamin B12 and choline chloride gives conflicting results. The avoidance of heat stress and moulded forages could be also helpful.

Slipped Tendon (Perosis)

Perosis or chondrodystrophy is encountered in young birds whose diet is deficient in manganese (Mn) or some of the following vitamins: choline, nicotic acid, pyridoxine, biotin or folic acid. This is an anatomic deformation of leg bones in young chickens, turkey poults, pheasant poults, etc. It is characterised by retarded growth of long bones, widening of the tibiometatarsal joint, twisting or bending of the distal end of tibia and the proximal end of metatarsus and finally, slipping of the gastrocnemius muscle tendon from its condyles. Clinically, it is manifested by impaired locomotion because of leg lateral and posterior malposition of the leg.

The vitamin D3 deficiency or lack, or the impaired ratio of calcium and phosphorus result in rachitis in young birds. It is encountered more frequently in industrial poultry breeding where the forages are often inadequately balanced. The deficiency of vitamin D3 and phosphorus are more common in growing birds whereas the calcium deficiency in young or adult layer hens. The newly hatched chickens have depleted calcium stores. The calcium deficiency is rapidly manifested if they do not receive an adequate diet supplement. Clinically and morphologically, soft bones or various degree of bone deformation are determined. The birds usually lie down and their growth is retarded. The diagnosis is based upon the complex evaluation of their age, signs and lesions. The supplementation of vitamin D3 with water or forage and the balancing of Ca/P ratio in the diet contribute to the favourable outcome of the disease.

Pulmonary Hypertension (Ascitis) Syndrome in Broiler Chickens

The pulmonary hypertension syndrome (PHS) or ascitis is related to the rapid growth and enhanced metabolic processes in broilers. Affected chickens are with a strongly distended abdomen, reluctance to move, respiratory troubles and cyanosis.

The pleuroperitoneal cavity of affected chickens is filled with straw-yellow fluid. The rapid growth in contemporary broilers is related to higher needs for oxygen, and the lung remains relatively small vs body dimensions. The main factor for PHS is hypoxaemia that results in enhanced cardiac activity. The consequences are pulmonary hypertension, right cardiac failure and ascitis.

Immediate causes for hypoxaemia are the inadequate ventilation, the high ammonia or dust levels in premises, low temperatures, stimulating the metabolic processes. Generalised passive venous hyperaemia and stasis in the liver and other viscera are observed.

The poseterior vena cava and the right half of the heart are over filled with blood. In advanced stages, the regenerative reparative processes in the liver could occur. The lypoxia favours the growth of connective tissue in the interstitium that results in thickening of the organ and appearance of sclerotic lesions.

The stasis is sometimes causing liver rupture and internal bleeding.

In a number of cases, a secondary contamination with E. coli could occur and therefore a serious fibrnous inflammation, manifested as polyserostis, could be observed.

A left compensatory hypertrophy is observed (transverse cross-section of a fixed heart). The right ventricle at an early stage is hypertrophied whereas at a later -dilated with extremely thinned wall (transverse cross sections of fixed heart). The diagnosis is based on gross lesions. The reduction of oxygen requirements by slowing down the metabolism protects from and decreases the incidence of ascitis. Restriction of forage consumption and various light patterns are recommended.

Amyloidosis

Amyloidosis is seen mainly in adult birds. It is characterised by extracellular buildup of the protein amyloid in different viscera and the joints. Amyloid arthropathy. It is primarily associated with Enterococcus faecalis and Mycoplasma synoviae. Brown layers are particularly susceptible. Some genetic factors could be involved in the incidence of amyloidosis too. The affected joints are enlarged, swollen and contain orange-yellowish matter. Amyloidosis of internal organs. The buildup is observed mainly in the liver, the spleen and the kidneys. The affected organ is multifold enlarged, with stretched capsule, rounded margins and pale colour. The state is usually seen following severe disturbances in protein metabolisms prolonged and exhausting diseases (tuberculosis, etc.).

The cannibalism in birds is usually consequence of feather pecking or pulling out, or vent pecking, that are behavioural reactions. Feather pecking or pulling out are observed in birds, reared in closed, often overcrowded premises. This happens frequently in growing broiler breeder flocks during the period of restricted feeding.

The vent pecking is usually taking place soon after the beginning of egg-laying and could be related to occurring hormonal alterations.

Pecking of the skin in various parts of the body and the vent result in blood loss, protrusion of viscera from body cavities and death. Some of the states that are supposed to provoke feather pecking are the high light intensity in the premise, pelleted feed, nutritional and mineral deficiencies and skin injury by ectoparasites.

Formation of crusts after pecking in the head area in a turkey poult. The prevention of cannibalism could be done by providing adequate diets (ground feeds are preferred), reduced light intensity, avoiding overcrowding. An extreme measure is debeaking. Marked anaemic appearance of the carcass due to massive blood loss subsequent to cannibalism in a turkey poult.

Deep Pectoral Myopathy

The deep pectoral myopathy (DPM), called also green muscle disease or Oregon disease, is observed in heavy meat types of turkeys or chickens. The disease occurs because of ischaemic necrosis due to inadequate blood supply of variously sized deep pectoral muscle groups. The lesion is uni- or bilateral and is detected as a slaughterhouse finding. Affected muscles have an unusual green colour.

Histologically, the altered muscle fibres are enlarged at a various extent, markedly eosinophilic with rectic or lacking nuclei. At a more advanced stage, among the necrotised tissue, inflammatory reaction and replacement of atrophied fibres with fat or fibrous tissue could be seen.

Rupture of the Gastrocnemius Tendon in Broiler Breeders

This state could cause significant economical losses in broiler breeder flocks. It is usually seen in birds older than 12 weeks but is also observed in breeders at the age of 7 weeks or older than 24 weeks.

The rupture could be uni- or bilateral. Clinically, lameness is observed. In affected birds, swelling of the posterior surface of the leg, just above the tibiotarsal joint could be palpated.

In acute lesions, haemorrhages are visible through the skin. In older lesions, a blue-greenish discolouration is present whereas in chronic lesions, very hard masses as abnormal subcutaneous tissue are palpated.

In part of birds, the muscles of affected legs undergoes a various degree of atrophy. In acute lesions, swellings and haematomas under the skin of the posterior surface of the leg are detected.

The rupture is usually seen as irregular, transverse interruption just above the tarsal joint. Among the haematoma, the free end of the tendon could be seen.

In older or chronic lesions, the blood is partially or completely reabsorbed. The end of ruptured tendon and the adjacent tissue are involved at a various extent by a newly grown fibrous tissue. RGT should be differentiated from reovirus-and MS infections, where the gross and histological lesions are with a marked inflammatory character.

Cage Layer Fatigue

Cage layer fatigue syndrome in birds is characterised by an inability to stand on their feet and fragile bones. It is mainly observed in young layer hens reared in batteries in the period of maximum egg-laying. Affected birds lie down and stopped eating. Egg shells become thin.

The calcium deficiency is layer hens results in initial removal of calcium from bones, to complete depletion of the medullary bone and thereafter, of the bone wall. The bones are strongly thinned and spontaneous fractures, especially of the tibia and the femur could occur.

Although the severe calcium deficiency is often a triggering factor, the aetiology of the syndrome seems to involve other, yet unknown factors.

The supplementation of calcium, phosphate and multivitamin preparations in the diet and rinking water, the regulation of avian population density into cages and ensuring adequate nutritional and drinking fronts are also contributing for the favourable outcome of the condition.

Thyperkeratosis

A local form of hyperkeratosis in an ostrich. Increased amount of keratin in the horny tissues - down, feathers and beak - is present, resulting in their coarse appearance. The state is linked to impaired metabolism of sulphur-containing amino acids, vitamin A deficiency, etc.

Gout

The gout is characterised by retention and buildup of urates in tissues. It is seen as two separate syndromes: visceral gout and articular gout. The visceral gout is manifested with deposition of urates in renal tubules and the serous coats of the heart, the liver, the mesentery, the air sacs or the peritoneum.

The urate deposits on serous coats resemble a chalky white dust. Visceral urate deposits are generally due to renal failure. Possible causes for this could be obstruction of ureters, renal damage or dehydration.

Most commonly, visceral gout following dehydration is observed in newly hatched chickens after overheating or a more prolonged stay in the hatchery. Visceral gout outbreaks are related to vitamin A deficiency, treatment with sodium bicarbonate, mycotoxicoses, etc.

Articular gout is characterised with periarticular urate deposits (tophi), especially around the joints of toes and the foot. The joints are enlarged and toes - malformed.

After opening of affected joints, the periarticular tissue is white. A white semi-liquid matter, due to urate deposits, could also be seen.

In chronic cases, urate precipitations could be observed in the trachea, the comb, wattles, etc.

Urolithiasis is an aetiologically unknown state, occurring primarily in cage layer hens, characterised by obstruction of one or both ureters with urates, atrophy of one or more renal lobes and a various degree of renal and visceral gout. A number of aetiological factors are related to this condition: protein excess, calcium excess (3% or more), sodium bi¬carbonate toxicity, mycotoxins (ochratoxin, etc.), vitamin A deficiency and nephrotropic strains of the infectious bronchitis virus.

The lower phosphorus levels (under 0.6%) are probably helping the manifestation of the disease. There are no specific clinical signs except for the depression and the weight loss. The death rate could increase and persist around 2-4% monthly during the productive period. The total mortality is heavily affected flocks could reach 50%.

Hyperandrogenism in Broiler Chickens

Hyperandrogenism in broilers is a condition, manifested with marked signs of masculinisation in chickens from both genders. The first signs could start very early after the age of 10-12 days. A very noticeable reddening of the comb and wattles, coarse feathering of the face, strong growth of nails and a highly aggressive behaviour in about 100% of birds are observed.

Facial wounds following manifestations of aggressive behaviour.

The peak of clinically manifested masculinisation traits is between the age of 20 and 30 weeks in all chickens. The serum testosterone is manifold higher. In 23-day-old broilers, average testosterone concentrations were 709 ng/dl vs 36 ng/dl in intact chickens at the same age. The mycotoxins following contamination of some of forage components with moulds are assumed to be a possible cause for this state, but yet, it is not clear if this or another factor is causing hyperandrogenism.

View of 35-day-old broiler chickens with signs of hyperandrogenism. All chickens in the premise are with strongly developed secondary sexual characteristics.

Of Cleavers (Galium Aparine) Seeds on Healthy Status and Production Traits in Broiler Chickens

Galium aparine, cleavers or stickywilly, is an annual weed in cereals and earthed-up crops, is an herbal plant belonging to the Galium genus (herbarium).

The seeds (fruit) are dry, round or kidney-shaped, with smooth, corrugated or granular surface, sometimes covered with hooked prickles, brown or grey colour and are notched on one side („nuts").

On the left - seeds with coat, on the right without coat. In the gizzard of chickens, the seeds become white with a contrasting black-coloured unipolar groove, similar to an eye. Most probably, that is why they are called „snake eye".

Purulent necrotic processes in the gizzard of a chicken. The health is influenced by consequences of mechanical obstruction because of seeds' accumulation in the gizzard.

A direct toxic effect of Galium aparine is also supposed (primarily hepato- and nephrotoxic), due to release of toxic glycosides and other toxic substances from the seeds.

Gizzard Impaction in Turkey Poults

The gizzard impaction could results in high mortality in turkey poults during the first three weeks of life. Affected turkey poults are dehydrated, with empty gut, but the gizzard is filled with hard fibrous masses.

In some cases, the fibrous masses enter the first part of the duodernum or the lower parts of the small intestines. the impaction is resulting from eating litter that gizzards cannot ingest. The prevention is aimed at control of litter eating.

Round Heart in Turkeys (Dilated Cardiomyopathy)

The aetiology is not known, but genetic factors or early viral myocarditis is supposed. The state is lethal in turkeys at the age of 1 - 4 weeks. Pathoanatomically, a severe dilated cardiomyopathy, often accompanied with ascitis, hydropericardium and congestion of other organs is detected.

Acute Selenium Intoxication

Acute selenium intoxication in birds is usually seen after multifold overdosing of preparations containing inorganic selenium with the aim of prevention or therapy. In the early hours after intake of toxic doses of selenium, a high death rate (up to 100%) and massive haemorrhages in the liver are seen.

In birds that survived the later stage of the selenium toxicosis, liver haemorrhages are outlined on the background of dystrophic necrobiotic changes.

Acute Propane-butane Intoxication

In cases of acute intoxication with propane butane (due to damage in heating appliances where propane-butane is used as energy source for heating), asphyxia, cyanosis of the featherless skin, pulmonary oedema and subcapsular haemorrhages in the liver are observed.

Stasis in the right half of the heart, overfilling of the posterior vena cava (a) and oedema of lungs (b) following overheating. When the temperature in the premise reaches 35°C, the cooling in birds begins by convection, conduction and radiation as they have no sweat glands. The excessive respiration results in oversaturation with CO_2 and then, in asphyxia.

Pasted vent is usually observed in diarrhoea, when the rectal masses are sticked to adjacent feathers. The observed plugs prevent the evacuation of the next faecal masses. This results in dilatation and obstipation of the rectum.

Spontaneous Rupture of the Caudial Renal Artery in Turkeys

A case of spontaneous rupture of the caudal renal artery in broiler turkeys is observed at the age of 8 - 12 weeks. Some birds in an excellent body condition die. The carcasses are anaemic. The necropsy shows that in some birds, the entire body cavity is filled with clotted blood and in others - massive subcapsular coagula from the side of the affected kidney are detected.

In all cases, unilateral rupture was observed, primarily of the left renal artery.

Subcutaneous Emphysema

Subcutaneous emphysema in the right inguinal region. It occurs after penetration of air or gases in the subcutaneous connective tissue. Usually, it results after air-sac tearing or skin injury.

Subcutaneous emphysema in the region of the head.

Gastro-intestinal Impaction

A sacciform blind pouch occurring after attachment of the intestinal wall to an unabsorbed yolk sac, causing impaction.

Invagination: Involvement of anterior parts of the small intestine, the mesentery and the tops of the caecum into the ileum. Usually, it occurs with a strong intestinal peristalsis, often following ingestion of feed after a restrictive feeding regimen.

Rotation of the small intestine on the longitudinal axis of the mesentery, resulting in venous stasis and necrosis of the intestinal wall.

Crop Impaction: It is caused by hard, fibrous feed or litter, whose accumulation results in impaction.

Crop Impaction: The retained content into the crop sometimes causes putrefactive necrotic processes, affecting the crop wall and the covering skin.

Malformations

Persistent Right Oviduct: If the regression of the right Mullerian duct, that is embryonally present in female chickens as oviduct, does not occur, a cystic dilatation of this structure is seen, whose size varies from a long cystic formation to a big sac, filled with fluid: a/ functional left oviduct b/ persistent right oviduct

Absense of eyes (anophthalmia) and cranial malformations are frequent malformations in newly hatched chickens.

Open Body Cavities: An additional skin formation (pouch) to the left laterocaudal abdominal wall, with part of the abdominal organs found within.

Polypodia: The occurrence of more legs than normal is often seen in broilers. Such birds frequently survive up to the end of the fattening period.

2

Importance of Water in Poultry Health

Water is the most critical nutrient required for the production of poultry, but its importance is often neglected by poultry raisers. It makes 55-75 per cent of the chicken's body and 65 per cent of an ego. It serves to soften and hydrolyze feedstuffs for digestion. As an important constituent of the bipod and lymph it serves as a carrier of digested food and waste products.

It serves to control body temperature by evaporation through air sacs and lungs. It is the main constituent of mucus which acts as a lubricant for the Joints and muscles. Besides being an important structural component, water is a necessary agent in all body processes, which are chemical reactions between substances in a liquid medium, of which water is the principle constituent. Although water is normally considered as H_2O, all natural waters generally contains various impurities and other contaminants. It is becoming apparent that certain chemical or microbiological contaminants, present in water have important direct and indirect effects on avian health.

Sources of Water Contamination

Sources of water contamination are mainly categorized into the following five groups:

1. Wastes of human and animal origin
2. Run off from agricultural land
3. Domestic sullages
4. Industrial wastes
5. Accidental pollution.

Routes of Water Contamination

Surface water horn rivers, ponds, lakes end canals i« most prone to contamination. In India most of the sewage water is dumped back, into the river or canal from which the water is often taken for drinking purposes. Besides, animals and humans bathe in it. All these factors contribute to high level of micro-organisms in water. Most commonly used subsurface water sources like wells, hand pumps and tuba wells gets; contaminated by sanitary land till and water clogging in the vicinity. Sometimes wells become contaminated with bacteria following tree roots and loose walls. The water distribution system is also prone to contamination. If there are leaks in the water supply pipes. Outside water seeps into the pipeline. Bad plumbing practices also contribute to contamination of water distribution system.

Nearly all poultry raisers use underground or overhead tanks to storewater. These tanks are usually not cleaned and disinfected regularly. Tank covers are often left open giving access to birds, droppings and other contamination. Consequently, these storage tanks become breeding grounds for micro organisms. Poor sanitation habits of workers also aggravate contamination. At the consumer end (poultry), the water troughs of the litter floor gets contaminated by litter, and secretion and excretion of birds. In channel type of waterier in cage system, water is contaminated with nasal discharge of birds. Subsequently the same contaminated, water is consumed by other birds.

Water Borne Diseases

Because of the essential role played by water in supporting chicken life it also has, if contaminated, great potential fer transmitting a wide variety of diseases and illnesses.

Water Elimination and Wet Litter Problems

Water is eliminated by the bird mainly through feaces and respiration. The amount of water eliminated by a bird depends on many factors. Birds consume and [eliminate more water in hot weather. Rations that are high in fiber, protein, minerals (Nacl, Mg. K. S) or toxins increase the water consumption and excretion through feaces. Birds suffering from diseases of intestinal tract evacuate watery faeces (Diarrhea). If the drinking water is contaminated with toxic elements and bacteria, water consumption and elimination seems to increase. Often, due to faulty design and management of water flowing system and drinkers, lot of water finds its way through seepage, leakage and spillage from waterier.

All these factors in turn raise the moisture content of the litter. This leads to severe health hazards and down grading. Increased litter moisture flavours the growth of molds and coccidiat oocysts. Further, it flavours the microbial breakdown of uric acid excreted by the birds, to ammonia. Hence high litter moisture leads to increased production and accumulation of ammonia in a poultry house, which causes. kerato-conjunctivitis. Respiratory affections, regression of bursa of fabric us, immunosuppression, poor performance, hock scabs and breast blisters in broilers and several other stress related problems.

Purification of Water

In order to reduce the microbial load, purification of water is done by various methods, like boiling, filtration, U. V. Irradiation, chemical purification etc. U. V. irradiation of water: U. V. irradiation of water helps to reduce the bacterial load but this method finds little application in poultry industry, because of high cost, no safeguard against recontamination and of photo reactivation of bacteria. The U. V. lamp used in this equipment has a definite life and with age the intensity of light is diminished leading to low.purification efficiency. Chemical purification: This is the most commonly used method Of water purification. In this process the destruction of bacteria is accomplished by chemical oxidation of the complete cell structure. The chemical purifiers are added to the water at such low levels that they are not harmful to birds but are very destructive for the bacteria and virus.

General Guidelines for Efficient Water Management

The poultry raisers should follow the following guidelines for efficient and successful water management to prevent health hazards in poultry:

1. Good quality drinking water must be made available to the birds at all times in adequate amounts.
2. The water source must be regularly sanitized to ensure; that the bacterial and fungal load is reduced.
3. Bacterial and fungal analysis of the water is to be done periodically to gauge the efficacy of the water sanitization programme.
4. At yearly intervals the pipelines should be flushed out with a concentrated chlorine solution under pressure. This will not only disinfect the pipeline but also eliminate the rust developing in the pipes and valves.

5. Seepage or leakage of water from waterier and pipelines inside the house must be prevented. Drinker height should be adjusted weekly.
6. Covers of the storage tanks should always be kept closed to prevent wild bird droppings and other contaminants.
7. Waterier should be cleaned with a scrub brush each day and the old water should be removed from the building.
8. Ensure that all the birds are drinking by providing adequate number of waterier.
9. Choosing appropriate delivery system and maintaining those systems are essential.
10. Functioning of automatic drinkers (nipples, bell type) should be monitored regularly.
11. Daily record of water consumption is very much essential. A sudden drop in water consumption is often the early warning of potential problems with flock. A large increase in water usage is usually associated with leakage problems.

Ammonia Problem in Poultry Houses Strategies for its Control

Modem intensive poultry fanning calls for careful management of in house environment as the gaseous end/by-products of various physiological and biochemical reactions involving birds and other organic materials, gets accumulated and create serious manage mental problems. Among them, ammonia (NH.) is the most frequently encountered toxic agent in poultry houses. It is the principal end product of microbial breakdown of litter uric acid and its concentrations inside poultry houses generally remain below a level of 25 ppm under good manage mental conditions.

However, factors like attempts of reducing ventilation to conserve heat during winter (Anderson et al; 1964), reuse of old litter Or use of same litter for a long time (Reece et al; 1979) and poor litter management often lead to accumulations of NH3 well above this level and concentrations up to 75-100 ppm are commonly found in many poultry houses (Charles and Payne, 1966 a). Higher levels of litter moisture (> 25-30%), pH (>8) and temperature and presence of favourable bacteria, fungi and yeast hasten uric acid breakdown and thus increase NH3 production (Ivos et al; 1966). Poor ventilation further aggravates the problem by allowing accumulation of the produced NH3. Cage houses, particularly having broiler breeders too develop high NH3 levels, if litter is not removed frequently. Presence of NH3 in air can be sensed by human beings from 25 ppm onwards and theconcentrations may be measured by using precision gas detector or by exposing indicator paper wetted with distilled water to atmosphere for 15 seconds. pH goes up from 6 to 11 as NH3 concentration rises from 0 to 100 ppm (Mourn etal; 1969).

Effects of Ammonia

At higher levels, NH3 is found to depress feed consumption, body weight gain, onset of sexual maturity and egg production. Petkov (1966) has reported as much as 9% drop in egg production upon 2 month exposure to toxic NH3 levels. Birds showed watery eyes, closed eyelids and rubbing of eyes with wings due to irritating nature of NH3. pH of egg albumen was observed to rise, facilitating deterioration of white while egg yolk developed a deep orange colour, downgrading the market value.

Birds exposed to higher NH3 levels developed kerato conjunctivitis (Bullis et al; 1950), air saculitis and showed reduced respiratory rate and increased susceptibility to coccidiosis and various respiratory diseases like New Castle, infectious bronchitis etc., due to damaged ciliary and epithelial lining of respiratory tract. Poor carcass quality in terms of increased breast blisters was reported in broilers by Charles and Payne (1966 a).

However, hatchability is reported to be affected only at very high experimental levels, which are not observed in poultry houses (Panigrahi, 1989). Control Various approaches have been adopted to reduce NH3 production with the basic aim of either to inhibit uric acid breakdown or to bind the released NH3.

Better Management: *the easy way out* Proper maintenance of liner in good condition by avoiding housing birds at high stocking densities, prevention of cake formation by regular raking and proper water management and provision of good ventilation to remove dust, odors and moisture from air will greatly help in preventing this problem to occur.

Chemical Treatment of Utter: Addition of certain chemicals to litter is the common practice to obtain immediate but time bound effect. Paraformaldehyde, a mixture of polyoxymethylene glycols, disintegrate slowly and release formaldehyde gas which will chemically combine with atmospheric ammonia to produce no odorous, nontoxic and white product called hexamethylenetetramine and water (walker, 1964). The antimicrobial action f formaldehyde has the added advantage of reducing microbial populations both in air and in litter. Various workers have added this compound in flake form to litter and observed satisfactory effect up to about 3 weeks at 3% (w/w) inclusion.

Hydrated lime and super phosphate were found to be effective at levels of about 5% (w/w) and Ikg/m 2, respectively. These compounds are widely used owing to their cheap cost and value as fertilizer. Phosphoric acid, when sprayed on litter @ 0.4 kg/m 3 greatly reduced litter pH and inhibited NH3 production up to 3 weeks (Reeceetal; 1979).

Natural zeolites (clinoptilolite, erionite etc.) have been extensively used in Japan and US. ClinoptiloUte, when used in litter @ 5 kg/m 2, gave encouraging results.

Similarly, bentonites addition to litter @ 2 kg/m 2 was found to decrease NH3 production. Volatile fatty acids like a mixture of 60% acetic and 40% prop ionic acids were used successfully up to 3% (w/w) in litter. Further, in acute conditions, addition of antibiotics like thiopeptin @100 mg/kg may be practiced to have immediate effect.

Dietary Treatment

The possible role of dietary zeolites in controlling NH3 production, in addition to their proven toxin adsorption capability and performance stimulating effect, need to be studied further in view of the scanty information, available. Similarly bentonites addition to diet may also be of some use.

Dietary addition of antibiotics like Zinc bacitaracin or Thiopeptin @100 mg/kg also was found to depress NH_3 production from litter the urease inhibiting effect of hydroxamic acid, observed in large animals offers another great scope for its practical use in poultry rations.

Extract of Yucca schidigera, a plant commonly seen in Mexico and US is presently used widely in manufacturing commercial preparations because of its known depressing effect on NH_3 production from litter.

Others like Sorbic acid, adipic acid, gention violet, calcium propionate, copper Sulphate, ferrous Sulphate, formic acid and benzoic acid, significantly reduced dietary microbial counts and therefore, their addition to diet may have some depressing effect on litter NH_3 production by lowering litter microbial counts.

Microbial Treatment of Litter

Inoculation of litter with certain genera of microorganisms that reduce production of mycotoxins like Scopulariopsis may be extended for minimizing the flora that takes part in NH_3 production.

Conclusions

Productive performance and disease resistance of chicken are drastically affected by high atmospheric NH_3 concentrations. This problem may be alleviated by treating diet or litter with certain compounds that hinder the production of NH_3 either directly or indirectly.

3

Integrated Disease Prevention Management in Poultry

Various aspects of poultry health maintenance and disease prevention are outlined. The present day health problems are often multi factorial and require special attention to avoid disastrous consequences. Health is a state of complete physical, mental and social well being of poultry, which is an important factor determining the profitability of poultry production. Disease represents a restraint to reproductive potential and yield, feed conversion efficiency and product quality. The causative factors of disease are classified as agent, host and environment. The interaction of these three factors is required to start disease in birds. Intensive health monitoring of elite primary breeders, GGPs, GPs is essential for effective prevention of vertically transmitted disea ses. Integrated disease prevention and control involves all the measures designed to prevent or reduce the incidence, prevalence and consequences of the disease. Broadly these measures include 1. Controlling the reservoir or source of infection. 2. Breaking the routes of transmission and 3. Manipulating the susceptible host. The practical use of vaccines to prevent or reduce disease losses in poultry is discussed. The objectives of vaccination, the types of vaccine, vaccination programme, combined vaccines and vaccination failures are presented in a non technical manner.

The concept of preventive medicine in poultry is now more important than ever, because the occurrence of disease will limit the performance of the bird and prevent it from achieving its maximum genetic potential. The present day poultry health problems are seldom unifactorial, on the contrary. They are very often multi factorial, requiring special attend on to avert disastrous consequences. These problems, which may not be frank disease hut results in sub and time consuming to overcome. It is much better to avoid those happenings in the first place through an integrated system of health maintenance and disease prevention. Successful prevention depends on knowledge of causation, dynamics of transmission, identification of risk factors and flocks at risk, availability of prophylactic or early detection and treatment measures, application of these measures to appropriate or flock and continuous evaluation procedures applied.

Structure of Poultry Industry and Disease Control

The basic structure of the poultry in-dustry can be regarded as a Breeding Pyramid. The apex of the pyramid is occupied by Elite (Primary) Breeders, Great Grand Parents (GGPs) and Grand Parents (GPs), followed by Parents in the middle and commercials at the bottom of the pyramid (Hunton, 1993). The major genetic manipulations take place at elite and GGPs level. This involves careful recording of the main production parameters and also, selection for desired traints. At this level intensive health monitoring is essential, so that diseases which can be vertically transmitted or which can depress performance are excluded.

At this stage, the pathogens detrimental to the production and health of birds are to be eradicated from the elite breeders to GGPs, GPs and Parents can produce devastating results in the commercial poultry industry as hundreds or thousands of commercial birds are descended from each bird of the elite breeder flocks. So the disease status, immune system and genotype of elite flocks have potentially profound effect on the health of commercial flocks. The need to protect elite and GGPs from catastrophic and vertically transmitted diseases dictates high levels of biosecurity.

Concept of Health and Disease

Health is one of the most important factors determining profitability of all systems of poultry production. It is defined as a state of being that is not merely absence of disease, but a state of complete physical, mental and social well being of poultry. Health and disease lie along a continuum and there is no single cut-off point. The lowest point on the heath disease spectrum is death and the highest point corresponds to perfect health. It is thus obvious that health fluctuates within a range of optimum well being to various levels of dysfunction, including the state of total dysfunction, i.e. death.

Health and Disease Spectrum:

1. Perfect Health
2. Better Health
3. Freedom from Disease
4. Unrecognized Disease
5. Mild Disease
6. Severe Disease
7. Death.

Health status not only influences the average performance of a flock but also its variability. The perfect health implies that a bird should be able to express as completely as possible the potentialities of its genetic heritage. This is possible only when a bird is allowed to live is healthy relationship with its environment an environment that transforms genetic potentialities into phenotypic realities.

The disease is Just the opposite of health i.e., any deviation from normal functioning or state of complete physical or mental wellbeing. The term "disease" literally means "without ease" (uneasiness) disease, the opposite of ease when something is wrong with body function. Disease represents a restraint to reproductive potential and yield, feed conversion efficiency and product quality.

The causative factors of disease may be classified as Agent (Pathogen), Host (Bird) and Environment (Jordan, 1996). These three factors are referred to epidemiological triad. The mere presence of agent, host and favourable environmental factors is not sufficient to start the disease in bird. What is required is an interaction of these three factors to initiate disease process in the bird. The agent, host and environment operating in combination, determines not only the onset of disease which may range from a single case to epidemics but also the distribution of disease in the flock.

Agent Factors:

1. *Biologic agents:* The biologic agent may include viruses, Chlamydia, rickettdiae, mycophasmas, bacteria, fungi and para-sites, These agents exhibit certain biological properties viz., a. Antigenic variation, b Variation En virulence, c. Variation in immunogenic, d. Infectivity (numbers necessary to establish infection in the host), e. Tropism, f. Survival within and outside the host and g. Resistance/Susceptibility to antimicrobial and disinfectants.

2. Nutritional agents
3. Physical agents
4. Chemical agents
5. Mechanicalagents.

Host Factors (Intrinsic) :

1. Age
2. Gender
3. Immune status
4. Stress (associated with point of lay or social stress).

Environment factors (Extrinsic):

1. Climate
   a) Season
   b) Ambient Temperature
   c) Humidity
2. Air Quality
   a) Dust
   b) Gases like NHs, C02. And CH4
   c) Free radicals.
3. Housing
   a) Direction
   b) Ventilation.
4. Lighting
   a) Intensity
   b) Colour
   c) Duration
5. Management
   a) Floor/Cage rearing
   b) All-in-all out
   c) Multiple ages
   d) Vaccination & Medication
   e) Debarking, Dubbing & Sexing
   f) Handling

g) Transportation
h) Hygiene and disinfection
i) Disposal of dead birds, litter and hatchery waste.

6. Nutrition
   a) Deficiency or excess of nutrients
   b) Water quality and supply
   c) Feed toxins
   d) Feed contaminants
   e) Change of diet
   f) Drugs-antimicrobials, growth promoters
7. Biological
   a) Insects
   b) Rodents
   c) Wild birds
   d) Opportunistic infectious agents
8. Social
   a) Population density
   b) Sex ratio.

Disease Transmission

Infectious diseases do not occur by chance. It is required for a harmful infection environment. The disease causing organisms are introduced or transmitted to a poultry flock in a variety of ways.

- Carriers within a flock
- Birds recently acquired from an out side flock
- Eggs from infected breeder flocks
- Human feet, hands and clothes
- Dust, feathers and manure on equipment and supplies, such as trucks, coops and egg filler flats
- Pet birds, predators, rodents, flies and insects
- Live or contaminated vaccines
- Contaminated feed/feed ingredients
- Water
- Air.

Disease Prevention and Control

Disease control involves all the measures designed to prevent or reduce the incidence, prevalence and consequences of the disease. The basic approach in controlling disease is to identify the weak points of diseases and break the weakest links in the chain of transmission (Pattison, 1993). Broadly these measures include: 1) controlling the reservoir or source of infection, 2) breaking the routes.

Controlling the Reservoirs and Source of Infection

If the first link in the chain of causation (i.e. the disease agent) is deemed to be the weakest fink, logically, the most desirable control measure would be to eliminate the reservoir source. This includes all the measures directed to reduce the quantity of the agent available for dissemination.

1. Early identification of reservoir/source of infection
2. Treatment of affected birds
3. Epidemiological investigation
4. Isolation (separation) of infected birds from healthy birds
5. Serological testing and removal of carriers
6. Hygienic disposal of dead birds, litter and hatchery waste
7. through cleaning, washing and disinfection of farm houses, equipment
8. Feed and Water quality control
9. Hygienic hatching of chicks.

Breaking Transmission Routes

A major aspect of communicable disease control relates to "breaking the chain of transmission". Microbes travel from place to place by chicken crates, egg filler flats, trucks, other equipment and people. They can be transmitted from one flock to the other through workers. The organisms are found on the skin and in the digestive and respiratory organs of dogs, cats, rats, mice and wild free flying birds, Flies, beetles and other insects are also notorious spreaders of disease causing organisms. Blocking the routes of transmission imply on control of environmental factors, that is, to bring about an adjusted equilibrium between host and environment through encouraging some ecological influences and inhibiting others (Woodger.1996).

- Human traffic control, such as locking the doors and banning the visitors.

- Providing foot dips (at the entrance of all houses), wheel dips at the entrance of the farm
- Aerial and water sanitation
- Pest control (flies, mosquitoes, ants, rodents, litter beetles, lice, bed bugs, wild birds, cockroaches, spiders, mites and weeds)
- All-in-all out farm management
- Restriction of movement of personnel, feed, vehicles and farm equipment
- Personnel hygiene
- Avoiding of on farm trading.

Manipulating the Susceptible Host (Birds at Risk)

The third link in the chain of transmission is the susceptible host or birds at risk. One effective way of controlling the spread of infection is to strengthen the host defense (Cheville, 1979). Under certain circumstances this may be accomplished by the following ways:

1. Genetic improvement of disease resistance by conventional selection and genetic engineering (gene, transfer).
2. Manipulation of immune system by
    a) Vaccination
    b) Avoiding social, environmental and managerial stress.
    c) Providing good quality, well balanced feed
    d) Immunomodulation.

Vaccination

Over decades of research, avian immune system has been successfully manipulated to benefit the poultry industry. The efficient functioning of avian immune system is judged through successful vaccination programme.

It is one of the most powerful and cost effective weapons of poultry preventive medicine. Diseases with both demonstrated and predicted potential for producing disastrous losses have been controlled through vaccination.

Vaccination is the second line of defense against disease losses, and it keeps disease losses within acceptable limits when sanitary management fails. Vaccination has to be designed according to the needs of the situation (Beard, 1979).

Vaccination objectives: The first key to develop an effective vaccination programme is to determine the vaccination objectives. The most important objectives of vaccination in poultry are as follows:

1. To prevent or reduce disease losses due to morbidity or mortality in vaccinates.
2. To prevent spread of diseases
3. To provide parental antibodies to chicks of immunized hens
4. To prevent adverse effects of vertically transmitted diseases in chicks
5. To prevent diseases which infect young chicks and subsequently results into development of unprofitable adults?
6. To protect against disease which damage immune system?
7. To reduce condemnation at slaughter
8. To prevent or reduce losses due to poor egg quality.

Types of Vaccines: Once the objectives are established and decision is made to pursue the vaccination pro-gramme, the type of vaccine, frequency/timing and route of administration should be matched to those objectives. Vaccines for poultry are of two types, either live or killed, each having specific advantages or uses. Live vaccines contain a virus or bacteria that must infect the chicken and multi-ply in its body to produce immunity, hope fully with minimal reaction. Advantages of live vaccines are ease of administration, low price, rapid onset of immunity and a broader scope of protection, since chickens are exposed to all stages of the replication virus. Killed vaccine is prepared from bacteria or virus which have been inactivated and processed. Killed vaccines are usually combined with adjuvant such as aluminum hydroxide or oil. Adjuvant enhance the immune response by increasing the stability of vaccine in the body which stimulates the immune system for a longer period of time. Advantages of Rifled vaccines are assurance of administration of a uniform dose, safety, development of uniform levels of immunity and increased product stability.

Vaccination Programme: Vaccination programme vary with the type of poultry and severity of the field challenge. The immune responses needed in broilers are for a short duration because of the short life span of the bird (six weeks). Breeders and layers require different programme because of longer life span (Aini, 1980). When designing a vaccination programme, one must first determine the disease challenges faced by the poultry industry in the region. To be effective, a vaccination challenge should be provided before any natural exposure to infection.

4

Epidemiology of Newcastle Disease and the Economics of its Control

This chapter is restricted to a consideration of Newcastle disease as it affects the birds that I refer to as village chickens, and that others regard as rural chickens or family chickens. I have been for many years project leader to programs funded by the Australian Centre for International Agricultural Research (ACIAR) and concerned with the development of Newcastle disease vaccines for use in village chickens. Of recent years there has been increasing interaction between the disparate organisations that have an interest in village chickens and the potential contribution of chickens to the alleviation of poverty amongst the rural poor. This group now includes FAO, IAEA, INFPD, World Bank, Danida and numerous NGOs. I welcome this opportunity to visit Denmark and meet with the Network for Poultry Production and Health in Developing Countries.

The view of ACIAR has been that little progress could be made with village poultry until Newcastle disease was controlled. Now that suitable vaccines are available it is possible to adopt a more holistic approach to the whole problem of rural chickens, and indeed of all rural poultry.

The Virus

The virus of Newcastle disease is classified within the genus *Paramyxovirus* of the family *Paramyxoviridae*. This immediately tells us certain immutable characteristics of the virus. The virus will have a genome of single stranded RNA. The inexact replication of the RNA will frequently produce variants with differences, often subtle differences, in phenotype from the parent particle. Unless there is suitable selection pressure, these variants will not prosper. We must be aware that the populations of Newcastle disease virus that spread in the field, or the populations that make up a vaccine stock, are not clonal. Selection pressure can alter the average behaviour of the population. Of particular interest to this discussion are the variations that can evolve in pathogenicity and in thermostability.

The infectious virus particle (the virion) will have a lipoprotein envelope that will be essential for infectivity. The proteins on the envelope will be specified by the viral genome. These will be antigenically important and they will contribute to the host specificity and the spectrum of pathogenicity of the virus.

We will also be able to suggest other properties of Newcastle disease virus, by biological analogy with other paramyxoviruses. In particular, we will expect that Newcastle disease virus will be antigenically stable across its geographical range, and with time. Although variants will be detectable with monoclonal antibodies, or by sequence analysis, polyvalent antiserum will not easily distinguish between strains.

Newcastle disease viruses are usually cultivated in the cells lining the allantoic cavity of embryonated hen eggs. Some strains kill the embryos; others do not. The virus will also grow in cell cultures of avian origin, and in some mammalian cells. The replication of some strains of the virus is indicated by the destruction of the host cells, a process termed cytopathogenicity. Not all strains of Newcastle disease virus are cytopathic and detection of these strains in cultured cells can be difficult.

All strains of Newcastle disease virus will agglutinate chicken red blood cells *in vitro* (and sometimes red blood cells from other species). The process is known as haemagglutination and is the basis of the common serological test, the haemagglutination-inhibition test, used to detect antibodies to this virus. Other serological tests are available.

Diseases as a Risk Factor in Relation to the Rural Poultry Model in Bangladesh

In addition to the question of breed quality of the chickens used in the SLDP, it has been underlined several times during evaluations of projects that the poor health status of the day-old chickens provided by the poultry model represents a major problem.

A number of people (avian pathologists associated the Department of Livestock Services' (DLS) poultry farms, directors of DLS farms, staff at Central Disease Investigation Laboratory (CDIL), Bangladesh Livestock Research Institute (BLRI), Bangladesh Agricultural University (BAU), the NGOs BRAC and PROSIKA, consultants to the private poultry sector, chicken- and key rearers at village level) have all expressed that little improvement of the health status of chickens at most levels of the production pyramid of the poultry model has been obtained during the most recent years.

Important poultry diseases for any production system such as Newcastle disease, fowl typhoid, pullorum disease, fowl cholera, Gumboro, Marek's disease and fowl pox have all been diagnosed in Bangladesh and are suspected to be responsible for a high percentage of the significant mortality reported.

Other infections, which are known to influence production results negatively such as mycoplasma infections, infectious bronchitis, coryza and parasites, are probably common.

A chicken rearer recently reported that the highest total mortality she had experienced in one of her batches of chickens (420) over a two-year period was approximately 36 % within the 8 week rearing period. Due to insufficient record and reporting systems no specific mortality rates from the government hatchery farms have been obtained but it is considered to be very high.

From time to time, the NGOs also experience high mortalities and disease outbreaks. In the private commercial sector where biosecurity has reached a higher level than at the government farms 15% mortality during rearing of the parent flocks is considered good.

It is estimated that the mortality at most of the government hatchery farms at times can be significantly higher. Not all mortality is due to disease but such mortality rates clearly indicate that the disease situation represents a major problem in the breeding and multiplication system and therefore also to the rural poultry model.

In this respect it can be mentioned that the diagnosis made often are based on a few uncertain criteria even at official laboratories. Newcastle disease, for instance, is often diagnosed only on the basis of bleedings of the alimentery tract.

As a result of this other significant forms of Newcastle may be overlooked. Consequently, no information about the real situation of Newcastle in Bangladesh is available.

Biosecurity and Management Aspects

DLS Farms

Presently, the buildings and infrastructure at the government farms are not found suitable for breeding and support of breeding stock and the personnel at the farms are not well aware of "risky procedures" and good management practices.

"Open" or "semi-closed" poultry houses used for different purposes such as laying, rearing and brooding are often located within a very small radius on the farms resulting in a high risk for transmission of diseases between houses. The same houses are constructed with openings in the walls covered with wire, which are almost impossible to clean and disinfect. The open houses are also very vulnerable to diseases, which are known to have a reservoir in the wild life. These include influenza, Newcastle disease, salmonella infections and fowl cholera. None of the houses have shower in facilities for the personnel.

Eggs and chickens of various ages are exchanged quite intensively between the different DLS farms and represent a serious risk of spreading of diseases. The reason for this exchange is the lack of an overall production and multiplication plan.

The staff at the farm goes in and out of the different houses without any precautions and no fixed procedures have been implemented. Simple precautions such as the use of foot-dips or change of clothes etc. at the entrance to all houses are not taken.

Vaccines (primarily Gumboro) are still not always available. When the farms run out of vaccine stock the application procedure for supply of new batches of vaccines may take a long time with a resulting temporary lack of vaccine. Another problem is that efficacy testing has not been carried out on vaccines produced by DLS and no consistent vaccination policy has been made. Vaccination against Marek's disease is only performed in case of acute problems. But the diagnosis of Marek is based on insufficient pathological criteria, which may result in wrong conclusions. Storage and use of vaccines are not performed strictly according to the manufactures' recommendations.

No serological or bacterial routine monitoring of the flocks are performed before the start of laying or during the laying period. Routine post mortems of dead birds are not performed.

In addition, the layout of the hatcheries does not always comply to sound biosecurity standards, and the staff at the hatcheries are not aware of common hygienic precautions needed to be taken when handling hatching eggs, including proper disinfection and storage. The lack of disinfection of hatching eggs under the conditions existing in Bangladesh should be regarded as unacceptable.

NGO Poultry Farms

BRAC: The most recent farms build by BRAC are modern parent stock farms constructed according to high biosecurity standards. That is, closed, fully ventilated houses where the rearing unit is located at a reasonable distance from the production houses. Shower-in facilities are used. The associated hatcheries are also distant from the poultry units and have their own entrance shower-in facilities etc. The layout of the hatcheries is according to highest standards and planning has been done in collaboration with the manufacturer of the hatchers.

However, also semi closed production systems with a resulting minor level of biosecurity exist within the production system of BRAC (two farms).

Farm management has a high priority according to BRAC and internal courses are frequent. The farm managers meet at one of the BRAC centres once a month to coordinate working routines and standards. Staff is sent to the manufacturer of the setters and hatchers for training in hatchery management. Due to the lack of qualified local veterinarians within poultry production, BRAC has employed a veterinarian from abroad. He is visiting the farms, doing post mortems and trains the staff etc.

BRAC is interested in setting up a monitoring and control system, but does not have the possibility with the existing laboratory facilities. According to BRAC there is an urgent need for diagnostic laboratory capacity in Bangladesh and they are constructing their own laboratory, which is expected to be operational late summer 1999. In the meantime serology (mainly ELISA) is performed at a laboratory run by a private poultry company.

PROSIKA: Currently PROSIKA has two parent-stock farms in operation. One modern closed housing farm and another semi-closed farm. Both farms contain a mixed population of commercial hybrid parent-stock for broiler and layer production.

In 1999 two more closed housing system farms are expected to be fully operational. Both these new farms will be parent-stock farms with a mixed population of commercial hybrid parent-stock for broiler and layer production.

Training of farm managers is performed abroad by the breeding company supplying the parent-stock. Hatchery management training has to some extend been done in India. The overall impression of farm management/level of biosecurity etc. is that PROSIKA do take precaution but they to could improve on these issues.

This organisation has also expressed the need for extended diagnostic services.

Laboratory Back-up Facilities

The nucleus of "Veterinary Investigation Service in Bangladesh" is CDIL located in Dhaka, but also eight "Field Disease Investigation Laboratories" (FDIL´s) located at Mymensingh, Manikganj, Joypurhat, Barisal, Gaibanhda, Serajgonj, Sylhet and Feni are part of the system.

The main tasks of the CDIL are to examine specimens received from the field veterinarians, farmers, government and private livestock and poultry farms or any other source and to render diagnosis along with appropriate line of treatment and suitable control measures. The CDIL receives specimens mostly from rural and urban areas in and around Dhaka. It also receives specimens from the FDIL´s.

The technical knowledge and procedures required to perform a simple but still beneficial disease monitoring and control programme for the breeding flocks appear to be present at CDIL (mainly monitoring the salmonella-, mycoplasma-, Newcastle disease virus status of the breeding flocks). However, it is impossible for the present number of scientific staff at CDIL to run the number of test required in a monitoring and control programme in parallel with their diagnostic duties. The other laboratories will have major problems in implementing the procedures needed. In general, there is a significant lack of chemicals, reagents, media etc. at these laboratories just to be able to solve the present diagnostic tasks.

The support, which would be needed from these laboratories in case of the introduction of a monitoring and control system of the breeding flocks, would further increase the demand for media and chemicals.

More detailed diagnosis and subsequent control measures cannot be obtained from any of these laboratories at present. Newer diagnostic tools such as commercial available ELISA tests are not available. As a consequence of this, one private poultry breeding company has started a serological laboratory where several ELISA tests are performed. BRAC is using this laboratory from time to time.

Future Actions Recommended to be Taken

Within the DLS production system, a new production setup is needed in order to be able to control the production as well as the disease situation. It is recommended that the DLS farm located at Savar operates as a central grandparent farm and the other DLS hatchery farms as parent-stock units. Attention should be given to the fact that housing of birds of different ages on the same site increase the risk of persistence and cross-infection and reduces or prevents the opportunity of achieving a total depopulation and proper cleaning of the site. On a multi-age farm it is also more difficult to maintain effective live vaccination programmes.

It is recognised that the DLS hatchery farms probably cannot be run as all-in, all-out sites, but due to the high disease pressure which is suspected at these government farms it is strongly recommended that the sites as a minimum are depopulated and cleaned and disinfected prior to the arrival of stock within a new production structure.

This could be done for one farm at a time. As soon as possible after a house has been depopulated, dry cleaning followed by wet cleaning and disinfection, and sanitisation of the drinking- and feeding systems according to a written protocol should be performed.

Traffic to and on farm and general hygiene, handling of equipment, rodent and pest control, hatchery management and hygiene, feed storage and vaccination are all important aspects, which need to be addressed by DLS, in particular.

The NGOs and DLS should work together in setting up health monitoring and disease control programmes of breeding flocks which initially should seek to eradicate pullorum disease, fowl typhoid, mycoplasma infections, Newcastle disease, Gumboro and leucosis from the top of the production pyramid.

Training of personnel at all levels of the production (including hatchery operation) is necessary to be able to achieve significant improvements.

Upgrading of laboratory back-up facilities must have a high priority with special emphasis on increasing the number of scientific staff.

Proper use of vaccines and consequent use of vaccination programmes is like-wise needed and require the availability of the relevant and efficient vaccines.

Research in the significance, prevalence, epidemiology etc. of diseases should be initiated as recommended in the "Applied research and study component poultry development" (Darudec, 1998).

In the following, a more detailed outline of important aspects to be addressed will be given.

Traffic and Hygiene

Traffic onto the farm is a major source of disease introduction, and humans are the single most important factor in the form of e.g. truck drivers, sales people and service crew who routinely go from farm to farm and flock to flock in a short period of time. The following can control human traffic at relevant farms:

- Keep all nonessential personnel off the farm.
- Establish a logbook and record non-farm staff that enters the farm
- It should be made possible to distinguish "dirty" and "clean" areas on the farm
- Confine employee parking to the dirty side of the farm
- Lock all doors and use gates to prevent unwanted vehicle traffic.
- Appropriate procedures should be taken when crossing from dirty to clean area, such as showering or change clothing and footwear.

In addition, it is strongly recommended that descriptions of the following routine operations at the farms are available for the staff and that they are understood:

- Egg collection
- Collection of spent hens
- Arrival of new stock

Equipment

Equipment used in handling, managing and moving birds, feed, eggs or manure is capable of mechanically transmitting diseases from one location to another. Movement of equipment from one farm to another should consequently be avoided. Trucks, particularly load-out trucks and associated equipment, can be heavily contaminated and should if possible be avoided on the farm. If they have to enter they should be carefully washed and disinfected before entering the farm. Rodents and other pests

Wild birds and other animals are significant sources of poultry diseases such as avian influenza, Newcastle disease and fowl cholera. Rodents and different beetles may be important sources of salmonella, *E.coli*, infectious bursal disease (Gumboro) and fowl cholera. Flies can transmit Marek's disease and intestinal worms. The following measures should be used to reduce these pests on the farm.

- Implement rodent and insect control programmes
- Screen and eliminate nearby nesting areas for wild birds
- Keep the vegetation round the houses down
- Keep litter as dry as possible to prevent flies from laying
- Eliminate all access points into the houses (this require of course closed housing system)

Hatchery Management Considerations to be Made

Location. Should be located at a distance from the poultry farm. (This is not always the case at the DLS hatchery farms).

Layout. Careful planning of the working areas of the hatchery can greatly assist in allowing hygienic work practices by separating "clean" and "dirty" areas.

Ventilation. Should be operated to support separation of clean and dirty areas.

Waste disposal. Proper arrangements for the disposal of hatchery waste will reduce the risk of contaminating nearby poultry units.

Site security. As with farms, proper security and control of visitors is essential cleaning and disinfection procedures of site as well as equipment.

Hand hygiene :of staff is important throughout the various stages of handling hatching eggs

Written procedures: for most working routines including collection of eggs at the farms.

Hatching & Hatchery Hygiene

The egg laid, is vulnerable to damage and bacteria and should be collected quickly. In that respect the number of floor eggs should be reduced by proper actions such as to ensure a sufficient number of nests placed optimally in the houses. It has been observed that this could be improved at the government hatchery farms. Production planning is important to be able to produce DOC in right numbers at the right time. Better timing and planning could be obtained in Bangladesh if better analysis and arrangements were made in collaboration between the NGO's and the government farms. God hatching eggs must:

- be fresh (stored less than 7 days)
- be stored under optimal conditions (15-20°C, 75% relative humidity)
- be properly disinfected as soon as possible following collection. This is very important under Bangladesh conditions with the high temperatures and high humidity experienced. Growth of fungi has been reported to be a significant problem at the hatcheries causing poor hatchability, contamination of newly hatched chicks with subsequent high mortality.
- be free of transmissible pathogens (such as mycoplasmas, salmonella, leucosis virus etc.)
- have a uniform, sufficiently high level of maternal antibodies as ensured by proper vaccination.

Feed

Besides quality control of feedstuff, measures should be taken to prevent contamination of feed with pathogens. Food stores should be bird and rodent proofed to prevent contamination with e.g. Newcastle disease virus from wild birds and salmonella. Stores should be cleaned between flocks.

Monitoring of the Flock's Health Status

The following should be evaluated and recorded daily:

- Water and feed intake
- Mortality and morbidity and suspected causes
- Bird behaviour and appearance (beak, eyes, feathers, vent, legs and droppings)
- Presence of external parasites.

Considering the poultry health situation in Bangladesh it is strongly recommended that a disease control programme be implemented in a new production system. It should include the following aspects but must not be taken as a complete programme. The setting up of such a programme would require further analysis

Pullorum Disease and Fowl Typhoid

As these diseases are transmitted vertically to the progeny, it must be sought to eliminate these infections in top of the production pyramid. Consequently, repeatedly blood testing of breeding flocks and culling of reactors must be performed. All birds at the breeding farms should be tested twice between 16 weeks and point of lay and subjected to consecutive clear tests 1 month apart.

An annual clear test of 60 birds/flock should be initiated. Screening should also be performed at the hatchery farms. The need requires further analysis. The test used can be the rapid plate agglutination (RPA) test already used by CDIL.

Mycoplasmas

The microorganisms are also transmitted through the egg and elimination is the goal. Before the GP and P flocks come into lay in should be tested by rapid serum agglutination test (RSA) for *Mycoplasma gallisepticum* and *synoviae* infection. This test is already used by CDIL. A programme of eradication should be established and leucosis could be included.

Newcastle Disease

Proper vaccination programmes should be established based upon local experiences. Further analysis of the situation is required before final recommendations can be made. Such analysis may actually save vaccine. Today, some of the farms vaccinate once a month. This could be avoided by clarification of the problems.

Gumboro Disease

It would be most helpful in the understanding of the Gumboro problem to use the ELISA test in the breeding flocks to obtain information on the size of the problem. A proper vaccination programme should be implemented based upon management used and virulence of strains prevalent.

Vaccination

It should be kept in mind that the first step in a good vaccination programme is proper management including biosecurity. It further depends on:

- Storage condition of the vaccine, including temperature, time and time lag between preparation and utilisation
- Concentration of the vaccine
- Application method
- Age of the birds

- Immune status of the flock
- Disease pressure

A vaccination programme for the breeders should initially include:

- Marek
- Newcastle
- Gumboro
- Fowl pox
- (Cholera).

How to start this effectively and the extent of it depends upon:

- The establishment of a disease surveillance programme to make proper disease identification
- Detailed management and biosecurity analysis
- Logistic investigations enabling risk assessment.

It is recognised that this cannot be implemented all at once, but it should be aimed for in the future just as the use of other vaccines. In addition to the periodical lack of vaccine supply there does not seem to be efficacy control with vaccines produced by DLS including Newcastle vaccine. It is strongly recommended to set up a control scheme for these self-produced vaccines.

Another aspect to realise is that, under field conditions vaccinations alone is insufficient to bring about effective control of Newcastle disease and other diseases and must therefore be accompanied by good hygiene. In poorly managed, over-crowded, badly ventilated conditions with inevitable underlying bacterial infections even the mildest vaccine strains may produce severe disease.

A general Review on some Important Diseases in Free Range Chickens

Poultry production has undergone rapid changes during the past decades due to the introduction of modern intensive production methods, new breeds and improved biosecurity and preventive health measures. Moreover, the management methods place high demands on proper health, hygiene and require only a small, but very skilled labour force. In developing countries, however, adoption of this type of production has been limited due to the high inputs as listed above. The progress in industrial poultry production methods has thus had little effect on subsistence poultry production in the rural and periurban areas in developing countries. In these countries access to poultry meat and eggs depends on village-level poultry production. Although poultry production is considered as secondary to other agricultural production systems it has an important role in supplying villagers with additional income and high quality protein. This system provides valuable protein through a low input system, now representing 30% or more of the protein consumed (FAO, 1998). Almost all families in developing countries keep a chicken flock with an average size of about 10 adult chickens, varying from 5 to 50 animals.

The majority of these animals are kept in free range scavenging systems, where the birds scavenge around the house during daytime. Primitive housing of the birds during the night, however, is often seen. Supplementary feed consists mainly of household wastes, insects, larvae and seeds. Mortalities observed are in the range of 80 - 90% within the first year after hatching. For the same reasons the owners never include chicks when they refer to the flock size. The mortality is believed to be caused by mismanagement, lack of supplementary feeding, predators and diseases.

Little research has been published on rural poultry health, despite the fact that up to 80% of the poultry population in Africa and Asia is kept by the households as free range chickens. Although solid data have not been published, Newcastle Disease (ND) is regarded as the principle factor limiting rural poultry production in all African and Asian countries. ND may kill up to 80% of household poultry in Africa, but is not expected to account for the high early mortality rate according to the authors.

In addition, detailed epidemiology of the disease in the village situation is largely unknown (Yongolo, 1997). Furthermore, recent studies have shown that other diseases are present in scavenging poultry communities.

Since most of our knowledge relies on seroprevalence studies, solid longitudinal studies on causes of mortality are strongly needed to improve our knowledge of the prevalence and significance of the individual diseases under village conditions. The following data therefore mainly reflect experience obtained under backyard conditions in developed countries.

Diseases

According to Jordan et al. (1996) and Calnek et al. (1997) poultry diseases can be divided into five groups, namely bacterial, viral, fungal, parasitic and nutritional diseases. Only the diseases of expected importance under village conditions, e.g. those causing high mortality rates in chickens are mentioned in these tables.

As seen a wide variety of diseases are expected to occur under village conditions. Some of these diseases are age specific, whereas others are encountered in all age groups.

Table: *Important bacterial diseases in free range poultry and the age group where the disease is most often observed*

Disease	***Age group***
Escherichia coli	All ages, but mainly chicks
Salmonella spp.	All ages, but mainly chicks
Salmonella pullorum	Chicks < 3 weeks
Salmonella gallinarum	Growers, adults
Pasteurella multocida	Growers, adults
Haemophilus paragallinarum (Coryza)	Growers, adults
Clostridium perfringens	All ages, but mainly growers
Mycobacterium avium	Adults
Mycoplasma gallisepticum	All ages
Mycoplasma synoviae	All ages

Table: *Important virus diseases in free range poultry and the age group where the disease is most often observed.*

Disease	***Age group***
*Marek's disease	> 6 weeks
*Leucosis	Adults
Newcastle disease	Mainly growers and adults
Fowl Pox	All ages
Infectious Laryngotracheitis	Growers, adults
*Infectious Bursal Disease "Gumboro	< 8 weeks
*immunosuppressive disease	

Table: *Important fungal diseases in free range poultry and the age group where the disease is most often observed.*

Disease	***Age group***
Aspergilloses	Chicks
Mycotoxicoses	All ages

***Table:** Important parasitic diseases in free range poultry and the age group where the disease is most often observed.*

Disease	Age group
Coccidiosis	Chicks, growers, (adults)
Histomoniasis	1-3 months
Nematodes	All ages
Haemoparasites	Chicks, growers
Ectoparasites	Chicks, growers

***Table:** Important nutritional diseases in free range poultry and the age group where the disease is most often observed*

Disease	Age group
Vitamin A, D & E	Chicks, growers
Other vitamins, minerals and amino acids	Chicks, growers

Discussion

Approximately 80% of the world poultry population is kept as free range poultry. The free range poultry production system has also been designated as the "low input - low output" system (Pandey, 1992). Mortality in this system is in the range of 80 - 90% within the first year after hatching and is believed to be caused by mismanagement, lack of fresh water and supplementary feed, predators and diseases. Of these, diseases are believed to be the main limiting factor to the production of indigenous chickens (Aini, 1990). Among causes of early mortality nutritional diseases might be expected to dominate due to shortage of supplementary feed before and after hatch.

In addition, the quality of hatching eggs might be questioned under the climatic conditions present in these countries. Lack of vitamins and protein weaken the chicks and make them vulnerable to other diseases and predators. Diseases are also easily contracted under free range conditions due to scavenging habits. With an unconfined type of management, disease control is very difficult to carry out and is therefore rarely practised by the owners.

As mentioned earlier, Newcastle Disease is believed to be the most important disease in free range systems. During out-breaks of the disease up to 80% of the population may die. This, however, is dependant on different factors including the virulence of the strain causing the out-break (Alexander, 1997). A recent study in Nicaragua has, however, shown that in ND-immunised birds mortality is still high. The majority of the mortality is found in chicks up to 3-4 month of age. In this group up to 52.5% of the animals died due to other causes than ND. Similar studies in Mali by Wilson et al. (1987) have shown that chick mortality is in the range of 60% within the first 3 months after hatching.

A study in Morocco has revealed that up to 58% of the village chickens had antibodies against Salmonella gallinarum and S. pullorum. Similar findings were reported by (Adesiyun et al. 1984) from Nigeria. Chryosostome and his co-workers (1995) reported that 10% of the village chickens had antibodies against S. pullorum and that 62% had antibodies against Mycoplasma gallisepticum.

Furthermore, 65% of the animals had antibodies against ND. In Mauritania, Bell et al. (1990) found that 17.5% of the birds had antibodies against S. pullorum and that up to 46.2% of the birds had antibodies against Gumboro disease. In the same animals 7.5% had antibodies against ND. In Tanzania, Permin et al. (1999) examined 600 live chickens and found the presence of a range of diseases. All animals were parasitised with one or more (up to 14 species) species of endoparasites.

In total 29 different species were detected in the study. Furthermore, 65.7% of the animals were parasitised with Cnemidocoptes mutans, Dermanyssus gallinae and/or Echidnophaga gallinacea. The animals were also infected with a range of haemoparasites, the most common being Plasmodium juxtanucleare and Aegyptinella spp.

Antibodies against Newcastle disease was seen in 7.3%, against Salmonella enteriditis in 2.0%, against Salmonella gallinarum/pullorum in 52.7%, against Infectious Laryngotracheitis in 58.3% and against Gumboro disease in 42.3%. Similar studies have to the knowledge of the authors not been carried out in Asian countries. The significance of all these diseases, however, remains to be investigated. In addition, it should be noted here that a general trend for these studies is that they have only looked for antibodies against selected diseases.

Long-term cohort studies, examining the causes of death, have to date not been carried out in the free range production systems. Important knowledge on the proportion of the individual disease of the overall mortality is thus not known. It is consequently postulated that other diseases than ND are present in free range poultry production systems and that a successful development of this production system is only achieved when the exact causes of death is known. Since publications on disease prevalence based upon post mortem examinations are not expected to be accepted by existing international peer-reviewed journals the W.P.S.A. should be addressed to establish an international journal dealing specifically with problems relating to scavenging and family poultry. Establishing such a journal would ensure a rapid implementation of research results obtained.

"This is the first ever such atlas to be produced for Indian farmers and poultry professionals. Its objective is to provide a handy tool to farmers for on-the spot diagnosis of diseases, under field conditions.

Although a large number of laboratory tests, some sophisticated like ELISA and PCR, have been developed for the diagnosis of poultry diseases, they are either not available or are beyond the reach of most farmers. Moreover, by the time birds reach laboratory, they are mostly decomposed and unfit for postmortem examination and disease diagnosis. The delay in diagnosis allows the disease to inflict mortality and ruin farmer's economy. The key to disease control then is on-the spot diagnosis. It is with this objective that the colour atlas has been produced-to make readily available to poultry farmers a handy tool for immediate and accurate disease diagnosis under field conditions.

A large number of poultry diseases, including the more deadly like Ranikhet and Gumboro, leave their footprints in the dead bird. From these characteristic postmortem findings, it is possible to arrive at a correct diagnosis. The atlas provides 150 coloured photographs and covers a wide spectrum of poultry diseases. For important diseases, several photographs are given so that at no stage their diagnosis is missed. Also, wherever considered appropriate, pictures of live birds showing symptoms of specific diseases are given. In addition, each disease is accompanied by a brief description that highlights its salient features and also records author's field observations.

As this book is intended exclusively for disease diagnosis, it should be read alongside author's another book - "Poultry diseases - A guide for farmers and poultry professionals (2004) by J.L. Vegad" - to obtain information on other aspects of poultry diseases, namely, cause, transmission, symptoms, diagnosis, treatment, and control.

Although the book is written to meet requirements of poultry farmers, it will be equally useful to poultry consultants, diagnostic laboratories, and will acquaint the veterinary students with poultry diseases under field conditions."

The world's industry makes a significant contribution to the supply of dietary animal protein in both industrial and developing countries.

The structure of India's poultry industry varies from region to region. While Independent and relatively small-scale producers account for the bulk of production, integrated large scale producers do account for a growing share of output in some regions. Integrators include large regional firms that incorporate all aspects of production, including the raising of grandparent and parent flocks, rearing DOCs, contracting feed, providing veterinary services and whole selling.

Although the world's poultry industry has recorded unprecedented expansion during the past two decades, constraints associated with overproduction, environmental pollution and restraints to international trade are influencing the rate of expansion in various regions.

The major factors influencing the poultry industry are climate, availability and cost of land, capital and labour, proximity to markets, competitive efficiency of integrators, consumer satisfaction through safety, quality and flock welfare, environmental acceptability of large operations and successful implementation of international trade and above all, the price of feed ingredients determine the profitability of poultry production.

1. Guide to Commercial Poultry Production.
2. Improving backyard Poultry-keeping.
3. Knowledge-level among Poultry entrepreneurs on scientific layer Farming.
4. Issues and Economics of Poultry Production.
5. Fowl Play: The poultry industry's central Role in the bird flu crisis.
6. The use of Vaccination in Poultry Production.
7. Use of Herbal Plants in Poultry Health Management in the small-scale Commercial Farming.
8. Profitable Poultry: raising birds on pasture.
9. Biochemical and histopathological analysis of Aflatoxicosis in growing Hens Fed.
10. Patterns of change in Poultry Production Industry.
11. Exposure to dust and bioaerosols in Poultry Farming.
12. Guidelines for the Establishment and Operation of Poultry Farms.

13. Poultry Production Management and Bio-security Measures.
14. Environmental controlled Poultry Farm.
15. Poultry Processing opportunities.
16. Pastured Poultry Production: An Evaluation of its sustainability.
17. Strategy of household Poultry Development in coastal regions.
18. Small Commercial and family Poultry Production (A Case Study).
19. Organic Poultry Products are they the real deal.

Poultry farming is the practice of raising poultry, chickens, turkeys, ducks, and geese, as a subcategory of animal husbandry, for the purpose of farming meat or eggs for food. More than 50 billion chickens are reared annualy as a source of food, for both their meat and their eggs. Chickens farmed for meat are called broilers, whilst those farmed for eggs are called egg-laying hens. In total, for eggs are called egg-laying hens. In total, the UK alone consumers over 29 million eggs per day. Some lens can produce over 300 eggs a year. Chickens will naturally live for 6 or more years. After 12 month, the hen's productivity will start to decline. This is when most commercial laying hens are slaughtered.

Economic Raising of Poultry (Broiler, Layer, Ducks, Quail and Turkey)

Poultry Layer Farming

Poultry egg and meat are important sources of high quality proteins, minerals and vitamins to balance the human diet. Specially developed breeds of egg type chicken are now available with traits of quick growth and high feed conversion efficiency. Depending on the farm-size, layer (for eggs) farming can be main source of family income or can provide income and gainful employment to farmers throughout the year. Poultry manure has high fertilizer value and can be used for increasing yield of all crops

Scope for Layer Farming and its National Importance

Poultry industry which provides cheap source of animal protein has taken a quantum leap in the last three decades evolving from a near backyard practice to a venture of industrial promotion. Poultry is one of the fastest growing segments of the agricultural sector in India today.

While the production of agricultural crops has been rising at a rate of 1.5 to 2 percent per annum that of eggs has been rising at a rate of 8 percent per annum. India is on the world map as one of the top five egg producing countries with 55.6 billion eggs produced during 2008 (FAO).

The poultry sector in India has undergone a paradigm shift in structure and operation. This transformation has involved sizable investments in breeding, hatching, rearing and processing. Farmers in India have moved from rearing nondescript birds to rearing hybrids which ensures faster growth, good liveability, excellent feed conversion and high profits to the rearers. High quality chicks, equipment, vaccines and medicines are available.

Technically and professionally competent guidance is available to the farmers. The management practices have improved and disease and mortality incidences are reduced to a great extent. The industry has grown largely due to the initiative of private enterprise, minimal government intervention, considerable indigenous poultry genetic capabilities and adequate support from the complementary veterinary health, poultry feed, poultry equipment and poultry processing sectors. The industry has created direct and indirect employment for 3 million people.

Financial Assistance Available from Banks/NABARD

Loan from banks with refinance facility from NABARD is available for starting poultry farming.

For poultry farming schemes with very large outlays, detailed project reports will have to be prepared. Banks provide financial assistance for the following purposes :

a. For construction of brooder/grower and layer sheds, feed store, quarters etc.

b. For purchase of poultry equipment such as feeders, waterers, brooders etc.

c. For creating infrastructure items for supply of electricity, feed, water etc.

d. For purchase of day old chicks or ready to lay pullets.

e. For meeting working capital requirement in respect of feed, medicines and veterinary aid etc. for the first 5 to 6 months (i.e. till the stage of income generation).

The Cost of land is not considered for loan.

Scheme Formulation for Bank Loan

A scheme can be prepared by the beneficiary after consulting local technical persons of State Animal Husbandry/Veterinary department, Poultry Corporation or private commercial hatcheries. If possible, they should also visit the progressive layer farms in the area and discuss the profitability of farming. A good practical training and experience on a layer farm will be highly desirable, before starting a farm. Regular and constant demand for eggs and nearness of the farm to the market should be ensured.

The scheme should include information on land, water and electricity facility, marketing aspects, training facilities and experience of entrepreneurs and the type of assistance available from State government, poultry corporation, local hatcheries.

It should also include data on proposed capacity of the farm, total cost of the project, margin money to be provided by the beneficiary, requirement of bank loan, estimated annual expenditure, income and profit and the period for repayment of loan and interest. A format developed for project report preparation for a commercial layer farm is given in Annexure -I.

Appraisal of Project

After the scheme is submitted to the bank it is examined for technical feasibility and economic viability.

Technical Feasibility : This would briefly include :-

a. Suitability of climate and potentiality of the area

b. Availability of inputs such as chicks, feed, medicines etc.

c. Technical norms

d. Infrastructure available for veterinary aid, marketing, training and experience of the beneficiary.

Financial Viability : This would briefly include :-

a. Unit cost and loan requirement.
b. Input cost for chicks, feed, veterinary aid, labour and other overheads.
c. Output cost i.e. sale of eggs, culled birds, for meat, manure, empty gunny bags etc.
d. Income-expenditure statement and annual gross surplus.
e. Cash flow analysis.
f. Repayment schedule i.e. repayment of principal loan amount and interest.

Other documents such as loan application form, security aspects, margin money requirement etc. are also examined. A field visit to scheme area is undertaken for conducting technoeconomic feasibility study for appraisal of the scheme.

Sanction of Bank Loan and its Disbursement

After ensuring technical feasibility and financial viability, the scheme is sanctioned by the bank. The loan is disbursed in kind in 2 or 3 stages against the creation of specific assets such as construction of sheds, purchase of equipment and machinery, recurring cost during growing period on purchase of chicks, feed, medicines and vaccines, electricity and water, labour expenses etc. for first cycle. Constant follow up and supervision of the scheme is done by the bank.

Lending Terms - general

Financial Outlay

Financial outlay of the project depends on the local conditions, unit size and the components included in the project. Prevailing market prices may be considered to arrive at the outlay.

Margin Money: Margin depends on the category of the borrowers and may range from 5 to 25% of the total outlay.

Interest Rate for ultimate borrower : Banks are free to decide the rates of interest within the overall guidelines. However, for working out the financial viability and bankability of the model projects we have assumed the rate of interest as 12 % p.a.

Security

Security will be as per NABARD/RBI guidelines issued from time to time.

Repayment Period of Loan

Repayment period depends upon the gross surplus from the project. The loan will be repaid in suitable monthly/quarterly instalments usually within a period of seven to nine years with first year as grace period.

Insurance

The birds and other assets (poultry sheds, equipments) may be insured. Wherever necessary, risk/mortality fund may be considered in lieu of insurance.

List of Medicine, Vaccine and Nutrient Manufacturers

Factors in Planning A Poultry Operation

Independent or Contract Production :

1. Independent egg producers have the total responsibility for all planning, producing and marketing of eggs. They must provide houses, equipment, birds, feed, all supplies and management. Their risks are greater but potential profit or loss is also greater. There are no independent broiler producers.
2. Contract poultry producers (broilers and eggs) provide land, labour, houses, equipment, taxes, utilities, and insurance. The contractor furnishes birds, feed, medication, supervision, and markets the product. The contractor picks up eggs or broilers at the farm and the producer is paid a base price per dozen or per pound plus an incentive bonus as stated in the contract.

In planning for a poultry operation, a determination should be made of all contractors operating within the area and the type of contract available.

Location within Florida

Egg farms are located in counties all over the state; however, the major egg producing areas are the Tampa Bay area, all across Central Florida, northward to Nassau County, and scattered independent producers located across the northern part of the state.

Independent egg farms could be located in any county, however, make sure a market is available and zoning laws permit poultry farming. It would be well, however, to consider locating new farms well away from populated communities.

All commercial broiler producers are located in North Florida (counties north of Gainesville).

It would not be feasible to establish a broiler farm in other counties, since essentially all broilers are produced under contract and, at present, contractors only operate in North Florida.

Investment

With building and equipment costs continuing to rise, it is difficult to set an accurate investment cost necessary to build and equip poultry houses.

Today's investment for buildings and equipment alone are running approximately $5.00 - $7.00 per bird for layers, depending on the degree of mechanisation.

Broiler houses and equipment will run approximately $3.50 per square foot. Broilers are usually housed at about 0.8 square feet per bird.

Independent egg farms may be any size. New egg farms usually have 50,000 layers per farm. Broiler farms have 25,000 - 90,000 bird capacity per farm.

Market

Since Florida is now an egg sufficient state, farms must produce eggs for a specific market. They can no longer produce eggs and then hope to sell them for a profit if they have to take a wholesale price for all eggs.

In planning for an egg operation, the most important consideration is to have a specific market or a firm contract from a producer. One of the biggest advantages which a farmer has in producing eggs under contract is the freedom from concerns of marketing.

Labour

The most successful poultry operations are those in which the owner and the owner's family are actively involved in managing and working on the farm.

Since poultry farming requires some work every day, the manager is responsible for labour seven days per week and most understand the need for this constant attention to details. The most successful poultry farmers enjoy this type of work.

Table 1: *Suggested Number of Layers per Man on Independent Farms Under Varying Conditions Marketing Type***

*Producer Type**	*A*	*B*	*C*	*D*	*E*
Type 1	13,000-17,000	9,000-11,000	7,000-9,000	6,000-8,000	5,000-7,000
Type 2	8,000-10,000	7,000-8,000	5,000-6,000	4,000-5,000	3,500-4,000
Type 3	7,000-8,000	6,000-7,000	4,000-6,000	3,500-4,000	3,000-3,500
Type 4	7,000-8,000	6,000-7,000	5,000-6,000	4,000-5,000	3,500-4,000
Type 5	6,000-7,000	5,000-6,000	4,000-5,000	3,500-4,000	3,000-3,500
Type 6	5,000-6,000	4,000-5,000	3,500-4,000	3,000-3,500	2,500-3,000

**Marketing Types as Follows:

A = Selling at wholesale with no grading involved

B = Selling at wholesale with grading being done at the farm

C = Semi-retailing (case lots) with no candling or cartoning

D = Semi-retailing (case lots) when candling and cartoning

E = Retailing door to door

*Producer Types as Follows:

1 = Producer who buys started pullets and whose operation is fully automatic - with pit cleaners, egg gathering belts, etc.

2 = Producer who buys started pullets and has automatic feeders

3 = Producer who buys started pullets

4 = Producer who raises his own pullets and is fully automatic with pit cleaners, egg gathering belts, etc.

5 = Producer who raises his own pullets and has automatic feeders

6 = Producer who raises his own pullets

Commercial Egg Production

Breeds and Strains Selection

1. Contract production - The contract egg producer usually has no voice in determining which strain of bird will be placed on the farm. The contractor will select the best bird for egg market.
2. Independent production - There are several top strains of egg production layers available. The producer should pick the strain to fit the market demand, such as egg size, egg shell colour, shell quality and feed efficiency. Chicks should be obtained from reputable hatcheries. By studying the Official Directory of Florida Breeding Flocks and Hatcheries the producer can find a source of quality chicks.

Housing and Equipment

Essentially all commercial layers in Florida are housed in cages. If an egg farm is going to produce eggs under contract, the producer will be able to get construction details from the contractor and often must build and equip the house following the contractor's directions. Since many contractor-owned complexes are built with closed sided, evaporative cooled houses, contract producers may be required to provide similar facilities.

Poultry House Essentials :

1. Locate on well drained land. High, sandy, black jack oak ridges make ideal locations.
2. Build on an elevated grade with good drainage between houses.
3. Consider acreage and location of buildings for future expansion.
4. Open houses should be built with ridge ventilation which can be opened and closed.
5. If houses are equipped with side curtains, the curtains should be adjustable.
6. Adequate house eaves (24-36 inches) to keep out blowing rain and direct sun are important.
7. Consider ease of manure clean out.
8. Make sure adequate land is available for manure disposal (approximately 1 ACRE/1000 birds).

Feeds and Feeding :

1. Feed and feeding instructions are provided by the contractors for farmers producing eggs under contract.
2. The independent producer should look at economics and buy high quality feed from a company offering good service as well as quality feed. Feeds are composed of important groups of ingredients called nutrients. A high quality feed will contain a proper balance of nutrients balanced for the purpose for which it is to be fed.
3. Most independent producers should start out buying complete feed. As he grows in size and gains experience, he may want to consider mixing his own feed. Investigate the economics, time and management necessary to mix a quality feed and maintain quality control. Details in formulation and feed mixing can be obtained from the Extension Poultrymen.

4. A feeding program should be developed to maintain quality. Check feed tanks for old, caked feed. Empty and clean tanks occasionally. Do not allow food to get wet.

Flock Health :

1. The contractor service person is trained to spot health problems before they develop into major problems and advise the producer.
2. The independent egg producer can best maintain healthy flocks by:
   (a) Using the all-in-all-out method of management.
   (b) Practising good management.
   (c) Keeping his farm and houses off limits to visitors.
   (d) Using a good vaccination program.
   (e) Becoming familiar with important diseases and diagnostic lab procedures.
3. Vaccination:
   (a) Marek's Disease: 1 day (at Hatchery)
   (b) Bronchitis: 10 days, 4 to 5 weeks, 14 to 16 weeks
   (c) Newcastle Disease: 2 weeks repeat every 2 to 3 months
   (d) Fowl Pox: High Risk Area - 1 to 2 weeks, 12 to 16 weeks; Low Risk Area - 8 to 16 weeks
   (e) Vaccination schedules may vary depending on the disease condition in the area.

Diagnostic Laboratories

There is a charge of diagnostic service. Birds may be sent to these labs by express or by person. Birds should not be shipped to arrive on weekends. Three or four live birds showing typical symptoms of the condition prevalent in the flock should be sent. The shipment should be accompanied by as much information concerning the flock and sick birds as possible such as: age, duration of illness, previous history of illness and vaccination, size of flock, rate of mortality, medication administration, feed intake, production patterns, etc. All birds submitted as well as containers will be disposed of at the lab.

Brooding

Floor houses :

1. Clean and disinfect houses and equipment completely.
2. Put in 4 to 6 inches of clean litre.

3. Have water and feed available when chicks arrive.
4. Inspect and adjust brooders prior to arrival of chicks.
5. Follow manufacturer's directions for brooder capacity.
6. Adjust brooder temperature at edge of hover at 95 degrees F. for the first week and reduce 5 degrees each week.
7. Use brooder guards for first 7-10 days.
8. Isolate brooder house and keep visitors out.

Cage Brooder houses

Many poultry farmers are now brooding pullets in cage brooders. Follow manufacturer's directions on heating systems. Use the same management recommendations for preparing house, feeding, watering and chick care.

Pullet Management :

1. Isolate pullet houses and keep visitors out.
2. Remove unthrifty pullets as they appear.
3. For feeding and lighting details check with an Extension Poultryman.
4. Weigh a sample of birds biweekly to check body weight against strain standard.
5. Debeaking - There are several satisfactory methods and programs of debeaking. The poultryman should select the program which fits his operation. The following methods are used:
    - Early Precision Debeaking
        - o Age of bird between 6 and 9 days.
        - o Both beaks inserted into a 10/64 or 11/64 inch hole.
        - o Hot blade (1,700 degrees F) cherry red.
        - o Hot blade is held against beak for 2.5 seconds.
        - o When done properly, one debeaking will last for the expected life of the bird.
    - TT Method
        - o Age of bird between 8 and 10 weeks.
        - o Sometimes used in addition to early precision.
    - Conventional Method
        - o Age of bird between 12 and 18 weeks.
        - o Independent block cutting of both beaks.

o Permanent and gives good protection during laying period.

Layer Management :

1. House only well developed, well fleshed pullets.
2. Use artificial lights to provide 15 hours of total light per day, and start with 1 foot candle intensity at the bird level.
3. Feed a balanced diet - Keep feed intake records.
4. Weigh a sample of birds monthly.
5. Remove obvious culls.
6. Observe birds often to spot problem areas early.

Egg Handling for Quality :

1. Keep cages and egg gathering equipment in good repair to prevent cracks.
2. Handle eggs carefully.
3. Clean dirty eggs using clean, warm (110-120 degrees F) water with a detergent sanitizer.
4. Cool eggs quickly and keep them cool. Hold at a temperature of 50-60 degrees F. with a relative humidity of 70-80%.

Hatching Egg Production and Hatchery Management

Flock Management

Keep birds healthy, use 10 males per 100 broiler breeder females or 8 males per 100 egg breeder females.

Feed

Feed a breeder diet which is specially formulated for high hatchability.

Care of Hatching Eggs :

1. Select eggs for size, shape, shell colour and texture, 23-28 oz. per dozen is best for hatching.
2. Hold eggs at a temperature between 55-65 degrees F., humidity of 70-85% and for not more than 7 to 10 days before setting.

Asia - Biomin again Hosted the Biennial Event, Biomin Asia Nutrition Forum

Biomin, a leading company in the global animal health and nutrition industry, again hosts the biennial event, Biomin Asia

Nutrition Forum. The forum catered specifically to the Asian audience, spanning over six cities across Asia from 10 to 21 October, from Cebu to Pattaya, Coimbatore, Chang Sha, Zheng Zhou and Tokyo. The theme of the forum is 'Sustainability: defining the basics, addressing the essentials, introducing NutriEconomics®'.

Professor David Farrell, a key speaker, said: "The great challenge in agricultural science today is to produce sufficient food in a sustainable system that will feed the nine billion people expected to be on this earth by 2050. As the price of food escalates and food riots are happening amongst the poor, we must get smarter to meet the challenge. There are solutions and there will be costs."

With more people to feed and finite resources to depend on, the feed and animal production industry has to grow profitably without compromising the ability to meet future needs. Targeting at key industry professionals from the poultry, pig, dairy and aquaculture sectors, the forum featured an impressive panel of leading industry experts like David Farrell, Professor from the University of Queensland; John Baize, President of John C. Baize and Associates; Robert Van Barneveld, Professor from the University of New England and Consultant Research Scientist (Nutrition) of Barneveld Nutrition Pty Ltd., and Maximilian Schuh, Consultant and experienced veterinarian in diseases of swine and cattle. Representing Biomin are Jan Vanbrabant, CEO of Biomin Asia; Guan Shu, Technical Manager of Biomin Asia; Jim He, Technical Nutrition Manager of Biomin Asia; Wolfgang Markert, Director Development Department of Biomin Holding GmbH and Franz Waxenecker, Director Innovation Management of Biomin Holding GmbH.

Delegates gained insights on various topics that address sustainability from different facets. Presentations explored the solutions to feed the growing population, define the basics of sustainability using the NutriEconomics programme and looked into the implications of commodity trade on animal production. In addition, there were discussions about feeding the livestock in 50 years, a review of the ban of AGP in Europe and identifying the expectations and trends of antibiotic growth promoter use in Asia. Biomin shared the latest mycotoxins survey results, revealed the future of mycotoxin degradation and unveiled innovations in animal health and nutrition.

5

Transmission and Infection of H5N1

Transmission and infection of H5N1 from infected avian sources to humans is a concern due to the global spread of H5N1 that constitutes a pandemic threat.

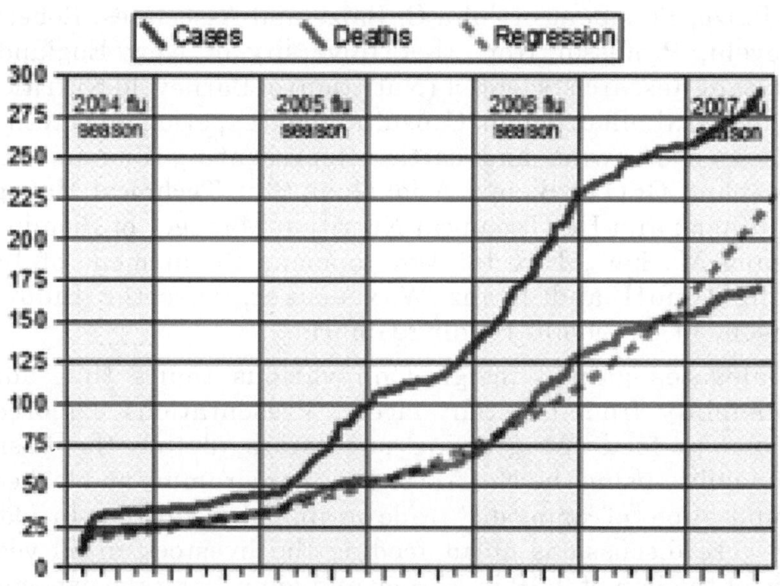

***Figure** : Cumulative Human Cases of and Deaths from H5N1*

Infected birds pass on H5N1 through their saliva, nasal secretions, and feces. Other birds may pick up the virus through direct contact with these excretions or when they have contact with surfaces contaminated with this material. Because migratory birds are among the carriers of the H5N1 virus it may spread to all parts of the world. Past outbreaks of avian flu have often originated in crowded conditions in southeast and east Asia, where humans, pigs, and poultry live in close quarters. In these conditions a virus is more likely to mutate into a form that more easily infects humans. A few isolated cases of suspected human to human transmission exist, with the latest such case in June 2006 (among members of a family in Sumatra). No pandemic strain of H5N1 has yet been found.

" The incidence of human cases peaked, in each of the three years in which cases have occurred, during the period roughly corresponding to winter and spring in the northern hemisphere. If this pattern continues, an upsurge in cases could be anticipated starting in late 2006 or early 2007." Avian influenza epidemiology of human H5N1 cases reported to WHO

The regression curve for deaths is $y = a + e^{kx}$, and is shown extended through the end of April, 2007.

H5N1 vaccines for chickens exist and are sometimes used, although there are many difficulties, and it's difficult to decide whether it helps more or hurts more. H5N1 pre-pandemic vaccines exist in quantities sufficient to inoculate a few million people and might be useful for priming to "boost the immune response to a different H5N1 vaccine tailor-made years later to thwart an emerging pandemic". H5N1 pandemic vaccines and technologies to rapidly create them are in the H5N1 clinical trials stage but can not be verified as useful until after there exists a pandemic strain.

Environmental Survival

Avian flu virus can last indefinitely at a temperature dozens of degrees below freezing, as is found in the northern most areas that migratory birds frequent.

Heat kills H5N1 (i.e. inactivates the virus).

Influenza A viruses can survive:

Over 30 days at 0°C (32.0°F) (over one month at freezing temperature)

6 days at 37°C (98.6°F) (one week at human body temperature) decades in permanently frozen lakes on hard nonporous surface such as plastic or stainless steel for 24–48 hours on clothes, paper and tissues for 8–12 hours.

While cooking poultry to 70°C (158°F) kills the H5N1 virus, it is recommended to cook meat to 74°C (165°F) to kill all foodborne pathogens.

Inactivation of the virus also occurs under the following conditions:

30 minutes 60°C (140.0°F) (half hour at a temperature that causes first and second degree burns in humans in ten seconds)

Acidic pH Conditions

Presence of oxidising agents such as sodium dodecyl sulfate, lipid solvents, and B-propiolactone.

Exposure to disinfectants: formalin, iodine compounds.

Ordinary levels of chlorine in tap water kill H5N1 in public water systems.

To kill avian flu viruses, the "World Health Organisation recommends that environmental surfaces be cleaned by the following:

Disinfectants such as sodium hypochloride, 1% in-use dilution, 5% solution to be diluted 1:5 in clean water, for materials contaminated with blood and body fluids.

Bleaching powder seven grams per litre with 70% available chlorine for toilets and bathrooms.

70% alcohol for smooth surfaces, tabletops, and other surfaces where bleach cannot be used"

H5N1 "can remain infectious in municipal landfills for almost 2 years. The two factors that most reduced influenza survival times were elevated temperature and acidic or alkaline pH."

Avian Flu in Birds

According to Avian Influenza by Timm C. Harder and Ortrud Werner

Following an incubation period of usually a few days (but rarely up to 21 days), depending upon the characteristics of the isolate, the dose of inoculum, the species, and age of the bird, the clinical presentation of avian influenza in birds is variable and symptoms are fairly unspecific. Therefore, a diagnosis solely based on the clinical presentation is impossible. The symptoms following infection with low pathogenic AIV may be as discrete as ruffled feathers, transient reductions in egg production or weight loss combined with a slight respiratory disease. Some LP strains such as certain Asian H9N2 lineages, adapted to efficient replication in poultry, may cause more prominent signs and also significant mortality. In its highly pathogenic form, the illness in chickens and turkeys is characterised by a sudden onset of severe symptoms and a mortality that can approach 100% within 48 hours.

The current method of prevention in animal populations is to destroy infected animals, as well as animals suspected of being infected. In southeast Asia, millions of domestic birds have been slaughtered to prevent the spread of the virus.

Poultry Farming Practices

There have been a number of farming practices that have changed in response to outbreaks of the H5N1 virus, including:

- vaccinating poultry against bird flu
- vaccinating poultry workers against human flu
- limiting travel in areas where H5N1 is found
- increasing farm hygiene
- educing contact between livestock and wild birds
- reducing open-air wet markets
- limiting workers contact with cock fighting
- reducing purchases of live fowl
- improving veterinary vaccine availability and cost.

For example, after nearly two years of using mainly culling to control the virus, the Vietnamese government in 2005 adopted a combination of mass poultry vaccination, disinfecting, culling, information campaigns and bans on live poultry in cities.

Dealing with Outbreaks

The majority of H5N1 flu cases have been reported in southeast and east Asia. Once an outbreak is detected, local authorities often order a mass slaughter of birds or animals affected. If this is done promptly, an outbreak of avian flu may be prevented.

However, the United Nations (UN) World Health Organisation (WHO) has expressed concern that not all countries are reporting outbreaks as completely as they should. China, for example, is known to have initially denied past outbreaks of severe acute respiratory syndrome (SARS) and HIV, although there have been some signs of improvement regarding its openness, particularly with regard to H5N1.

Use of Vaccines

Dr. Robert G. Webster et al. Write

Transmission of highly pathogenic H5N1 from domestic poultry back to migratory waterfowl in western China has increased the geographic spread.

The spread of H5N1 and its likely reintroduction to domestic poultry increase the need for good agricultural vaccines. In fact, the root cause of the continuing H5N1 pandemic threat may be the way the pathogenicity of H5N1 viruses is masked by cocirculating influenza viruses or bad agricultural vaccines."

Webster speculates that substandard vaccines may be preventing the expression of the disease in the birds but not stopping them from carrying or transmitting the virus through feces, or the virus from mutating.

In order to protect their poultry from death from H5N1, China reportedly made a vaccine based on reverse genetics produced with H5N1 antigens, that Dr. Wendy Barclay, a virologist at the University of Reading believes have generated up to six variations of H5N1.

Transmission

According to the United Nations FAO, wild water fowl likely plays a role in the avian influenza cycle and could be the initial source for AI viruses, which may be passed on through contact with resident water fowl or domestic poultry, particularly domestic ducks. A newly mutated virus could circulate within the domestic and possibly resident bird populations until highly pathogenic avian influenza (HPAI) arises. This new virus is pathogenic to poultry and possibly to the wild birds that it arose from.

Wild birds found to have been infected with HPAI were either sick or dead. This could possibly affect the ability of these birds to carry HPAI for long distances. However, the findings in Qinghai Lake-China, suggest that H5N1 viruses could possibly be transmitted between migratory birds.

Additionally, the new outbreaks of HPAI in poultry and wild birds in Russia, Kazakhstan, Western China and Mongolia may indicate that migratory birds probably act as carriers for the transport of HPAI over longer distances. Short distance transmission between farms, villages or contaminated local water bodies is likewise a distinct possibility.

The AI virus has adapted to the environment in ways such as using water for survival and to spread, and creating a reservoir (ducks) strictly tied to water. The water in turn influences movement, social behaviour and migration patterns of water bird species.

It is therefore of great importance to know the ecological strategy of influenza virus as well, in order to fully understand this disease and to control outbreaks when they occur. Most research is needed concerning HPAI viruses in wild birds. For example, small birds like sparrows, starlings and pigeons can be infected with deadly H5N1 strains and they can carry the virus from chicken house to chicken house causing massive epidemics among the chickens.

Avian Flu in Humans

Human to Human Transmission

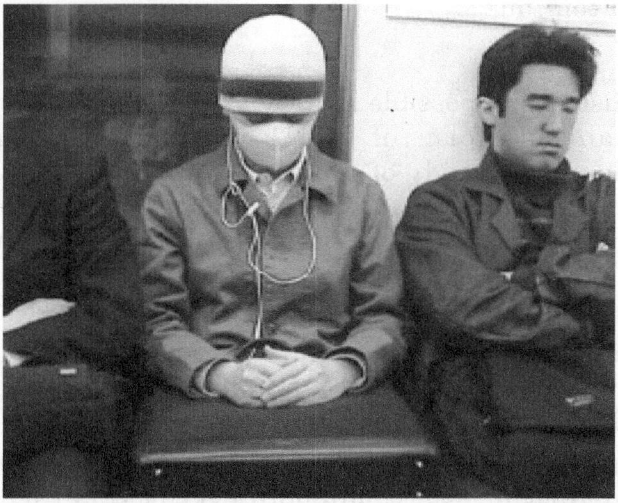

Figure: *The WHO believes that another influenza pandemic is as likely to occur at any time since 1968, when the last century's third of three pandemics took place. The WHO describes a series of six phases, starting with the inter-pandemic period, where there are no new influenza virus subtypes detected in humans, and progressing numerically to the pandemic period, where there is efficient and sustained human-to-human transmission of the virus in the general population. At the present moment, we are at phase 3 on the scale, meaning a new influenza virus subtype is causing disease in humans, but is not yet spreading efficiently and sustainably among humans.*

So far, H5N1 infections in humans are attributed to bird-to-human transmission of the virus in most cases. Until May 2006, the

WHO estimate of the number of human to human transmission had been "two or three cases". On May 24, 2006, Dr. Julie L. Gerberding, director of the United States Centres for Disease Control and Prevention in Atlanta, estimated that there had been "at least three." On May 30, Maria Cheng, a WHO spokeswoman, said there were "probably about half a dozen," but that no one "has got a solid number." A few isolated cases of suspected human to human transmission exist. with the latest such case in June 2006 (among members of a family in Sumatra). No pandemic strain of H5N1 has yet been found.

There is also concern, although no definitive proof, that other animals — particularly cats — may be able to act as a bridge between birds and humans. So far several cats have been confirmed to have died from H5N1 and the fact that cats have regular close contact with both birds and humans means monitoring of H5N1 in cats, and dogs will need to continue.

Prevention

Notwithstanding possible mutation of the virus, the probability of a "humanised" form of H5N1 emerging through genetic recombination in the body of a human co-infected with H5N1 and another influenza virus type (a process called reassortment) could be reduced by widespread seasonal influenza vaccination in the general population. It is not clear at this point whether vaccine production and immunisation could be stepped up sufficiently to meet this demand.

If an outbreak of pandemic flu does occur, its spread might be slowed by increasing hygiene in aircraft, and by examining airline cabin air filters for presence of H5N1 virus.

The American Centres for Disease Control and Prevention advises travellers to areas of Asia where outbreaks of H5N1 have occurred to avoid poultry farms and animals in live food markets. Travellers should also avoid surfaces that appear to be contaminated by feces from any kind of animal, especially poultry.

There are several H5N1 vaccines for several of the avian H5N1 varieties. H5N1 continually mutates rendering them, so far for humans, of little use. While there can be some cross-protection against related flu strains, the best protection would be from a vaccine specifically produced for any future pandemic flu virus strain. Dr. Daniel Lucey, co-director of the Biohazardous Threats and Emerging Diseases graduate program at Georgetown University has made this point, "There is no H5N1 pandemic so there can be no pandemic vaccine."

However, "pre-pandemic vaccines" have been created; are being refined and tested; and do have some promise both in furthering research and preparedness for the next pandemic. Vaccine manufacturing companies are being encouraged to increase capacity so that if a pandemic vaccine is needed, facilities will be available for rapid production of large amounts of a vaccine specific to a new pandemic strain.

It is not likely that use of antiviral drugs could prevent the evolution of a pandemic flu virus.

Symptoms

The human incubation period of avian influenza A (H5N1) is 2 to 17 days. Once infected, the virus can spread by cell-to-cell contact, bypassing receptors. So even if a strain is very hard to initially catch, once infected, it spreads rapidly within a body. For highly pathogenic H5N1 avian influenza in a human, "the time from the onset to presentation (median, 4 days) or to death (median, 9 to 10 days) has remained unchanged from 2003 through 2006."

Avian influenza HA bind alpha 2-3 sialic acid receptors while human influenza HA bind alpha 2-6 sialic acid receptors. Usually other differences also exist. There is as yet no human form of H5N1, so all humans who have caught it so far have caught avian H5N1.

Human flu symptoms usually include fever, cough, sore throat, muscle aches, conjunctivitis and, in severe cases, severe breathing problems and pneumonia that may be fatal.

The severity of the infection will depend to a large part on the state of the infected person's immune system and if the victim has been exposed to the strain before, and is therefore partially immune. No one knows if these or other symptoms will be the symptoms of a humanised H5N1 flu.

Highly pathogenic H5N1 avian influenza in a human appears to be far worse, killing over 50% of humans reported infected with the virus, although it is unknown how many cases (with milder symptoms) go unreported. In one case, a boy with H5N1 experienced diarrhea followed rapidly by a coma without developing respiratory or flu-like symptoms. As of February 2008, the "median age of patients with influenza A (H5N1) virus infection is approximately 18 years. The overall case fatality proportion is 61%. Handling of sick or dead poultry during the week before the onset of illness is the most commonly recognised risk factor. The primary pathologic process that causes death is fulminant viral pneumonia."

There have been studies of the levels of cytokines in humans infected by the H5N1 flu virus. Of particular concern is elevated levels of tumour necrosis factor-alpha (TNFá), a protein that is associated with tissue destruction at sites of infection and increased production of other cytokines. Flu virus-induced increases in the level of cytokines is also associated with flu symptoms including fever, chills, vomiting and headache. Tissue damage associated with pathogenic flu virus infection can ultimately result in death. The inflammatory cascade triggered by H5N1 has been called a 'cytokine storm' by some, because of what seems to be a positive feedback process of damage to the body resulting from immune system stimulation. H5N1 type flu virus induces higher levels of cytokines than the more common flu virus types such as H1N1 Other important mechanisms also exist "in the acquisition of virulence in avian influenza viruses" according to the CDC.

The NS1 protein of the highly pathogenic avian H5N1 viruses circulating in poultry and waterfowl in Southeast Asia is currently believed to be responsible for the enhanced proinflammatory cytokine response. H5N1 NS1 is characterised by a single amino acid change at position 92. By changing the amino acid from glutamic acid to aspartic acid, researchers were able to abrogate the effect of the H5N1 NS1. This single amino acid change in the NS1 gene greatly increased the pathogenicity of the H5N1 influenza virus.

In short, this one amino acid difference in the NS1 protein produced by the NS RNA molecule of the H5N1 virus is believed to be largely responsible for an increased pathogenicity (on top of the already increased pathogenicity of its hemagglutinin type which allows it to grow in organs other than lungs) that can manifest itself by causing a cytokine storm in a patient's body, often causing pneumonia and death.

Treatment

Neuraminidase inhibitors are a class of drugs that includes zanamivir and oseltamivir, the latter being licensed for prophylaxis treatment in the United Kingdom. Oseltamivir inhibits the influenza virus from spreading inside the user's body. It is marketed by Roche as *Tamiflu.* This drug has become a focus for some governments and organisations trying to be seen as making preparations for a possible H5N1 pandemic. In August 2005, Roche agreed to donate three million courses of Tamiflu to be deployed by the WHO to contain a pandemic in its region of origin. Although *Tamiflu* is patented, international law gives governments wide freedom to issue compulsory licenses for lifesaving drugs. A second class of drugs, which include amantadine and rimantadine, target the M2 protein, but have become ineffective against most strains of H5N1, due to their use in poultry in China in the 1990s, which created resistant strains.. However, recent data suggest that some strains of H5N1 are susceptible to the older drugs, which are inexpensive and widely available.

Research indicates that therapy to block one cytokine to lessen a cytokine storm in a patient may not be clinically beneficial.

Mortality Rate

From the first laboratory-confirmed case through March 18, 2008, the number confirmed human cases of H5N1 reported to WHO stands at 373, with 236 fatalities, reflecting a 63.27% fatality rate.

The global case fatality ratio looks only to the official tally of cases confirmed by the WHO. It takes no account of other cases, such as those appearing in press reports. Nor does it reflect any estimate of the global extent of mild, asymptomatic, or other cases which are undiagnosed, unreported by national governments to the WHO, or for any reason cannot be confirmed by the WHO. While the WHO's case count is clearly the most authoritative, these unavoidable limitations result in an unknown number of cases being omitted from it.

Human Mortality from H5N1

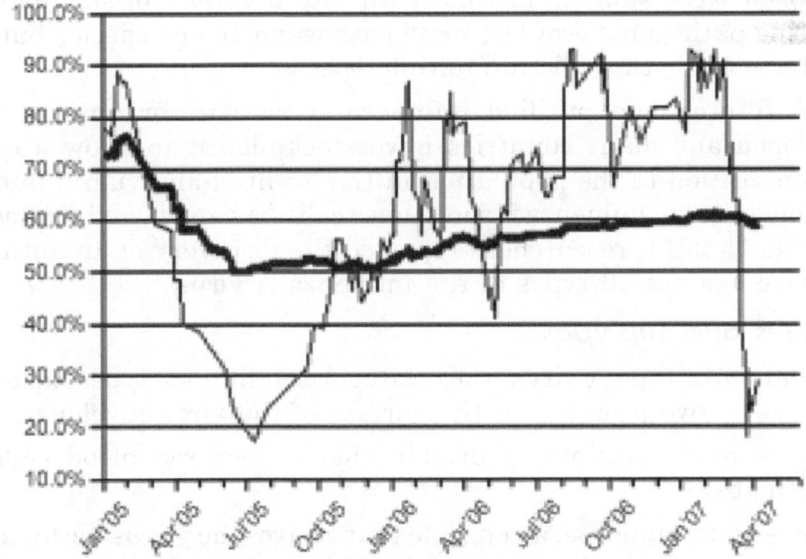

***Figure:** Source WHO Confirmed Human Cases of H5N1*

The thin line represents average mortality of recent cases. The thicker line represents mortality averaged over all cases. According to WHO: "Assessment of mortality rates and the time intervals between symptom onset and hospitalisation and between symptom onset and death suggests that the illness pattern has not changed substantially during the three years."

Influenza A Virus

Influenza A virus causes influenza in birds and some mammals and is the only species of Influenzavirus A. Influenzavirus A is a genus of the Orthomyxoviridae family of viruses. Strains of all subtypes of influenza A virus have been isolated from wild birds, although disease is uncommon. Some isolates of influenza A virus cause severe disease both in domestic poultry and, rarely, in humans. Occasionally viruses are transmitted from wild aquatic birds to domestic poultry and this may cause an outbreak or give rise to human influenza pandemics.

Influenza A viruses are negative sense, single-stranded, segmented RNA viruses. There are several subtypes, labelled according to an H number (for the type of hemagglutinin) and an N number (for the type of neuraminidase). There are 16 different H antigens (H1 to H16) and nine different N antigens (N1 to N9). The newest H type (H16) was isolated from black-headed gulls caught in Sweden and the Netherlands in 1999 and reported in the literature in 2005.

Each virus subtype has mutated into a variety of strains with differing pathogenic profiles; some pathogenic to one species but not others, some pathogenic to multiple species.

A filtered and purified influenza A vaccine for humans was developed and many countries have stockpiled it to allow a quick administration to the population in the event of an avian influenza pandemic. Avian influenza is sometimes called avian flu, and commonly bird flu. In 2011, researchers reported the discovery of an antibody effective against all types of the influenza A virus.

Variants and Subtypes

Influenza type A viruses are categorised into subtypes based on the type of two proteins on the surface of the viral envelope:

H = hemagglutinin, a protein that causes red blood cells to agglutinate.

N = neuraminidase, an enzyme that cleaves the glycosidic linkages of the monosaccharide, neuraminic acid

Different influenza viruses encode for different hemagglutinin and neuraminidase proteins. For example, the H5N1 virus designates an influenza A subtype that has a type 5 hemagglutinin (H) protein and a type 1 neuraminidase (N) protein. There are 16 known types of hemagglutinin and 9 known types of neuraminidase, so, in theory, 144 different combinations of these proteins are possible.

Some variants are identified and named according to the isolate that they are like and thus are presumed to share lineage (example Fujian flu virus like); according to their typical host (example human flu virus); according to their subtype (example H3N2); and according to their deadliness (example LP, Low Pathogenic). So a flu from a virus similar to the isolate A/Fujian/411/2002(H3N2) is called Fujian flu, human flu, and H3N2 flu.

Variants are sometimes named according to the species (host) the strain is endemic in or adapted to. The main variants named using this convention are:

- Bird flu
- Human flu
- Swine influenza
- Equine influenza
- Canine influenza.

Variants have also sometimes been named according to their deadliness in poultry, especially chickens:

- Low Pathogenic Avian Influenza (LPAI)
- Highly Pathogenic Avian Influenza (HPAI), also called: deadly flu or death flu

Most known strains are extinct strains. For example, the annual flu subtype H3N2 no longer contains the strain that caused the Hong Kong flu.

Annual Flu

The annual flu (also called "seasonal flu" or "human flu") in the U.S. "results in approximately 36,000 deaths and more than 200,000 hospitalisations each year. In addition to this human toll, influenza is annually responsible for a total cost of over $10 billion in the U.S.".

The annually updated trivalent influenza vaccine consists of hemagglutinin (HA) surface glycoprotein components from influenza H3N2, H1N1, and B influenza viruses.

Measured resistance to the standard antiviral drugs amantadine and rimantadine in H3N2 has increased from 1% in 1994 to 12% in 2003 to 91% in 2005.

> *"Contemporary human H3N2 influenza viruses are now endemic in pigs in southern China and can reassort with avian H5N1 viruses in this intermediate host."*

F16 Antibody

F16, an antibody that targets the hemagglutinin protein, was recently discovered and is the only known antibody effective against all 16 subtypes of the influenza A virus.

Structure and Genetics

> *"The physical structure of all influenza A viruses is similar. The virions or virus particles are enveloped and can be either spherical or filamentous in form. In clinical isolates that have undergone limited passages in eggs or tissue culture, there are more filamentous than spherical particles, whereas passaged laboratory strains consist mainly of spherical virions."*

The Influenza A virus genome is contained on eight single (non-paired) RNA strands that code for eleven proteins (HA, NA, NP, M1, M2, NS1, NEP, PA, PB1, PB1-F2, PB2). The total genome size is 13,588 bases. The segmented nature of the genome allows for the exchange of entire genes between different viral strains during cellular cohabitation. The eight RNA segments are:

HA encodes hemagglutinin (about 500 molecules of hemagglutinin are needed to make one virion) "The extent of infection into host organism is determined by HA. Influenza viruses bud from the apical surface of polarised epithelial cells (e.g. bronchial epithelial cells) into lumen of lungs and are therefore usually pneumotropic. The reason is that HA is cleaved by tryptase clara which is restricted to lungs. However HAs of H5 and H7 pantropic avian viruses subtypes can be cleaved by furin and subtilisin type enzymes, allowing the virus to grow in other organs than lungs." NA encodes neuraminidase (about 100 molecules of neuraminidase are needed to make one virion).

NP Encodes Nucleoprotein

M encodes two matrix proteins (the M1 and the M2) by using different reading frames from the same RNA segment (about 3000 matrix protein molecules are needed to make one virion).

NS encodes two distinct non-structural proteins (NS1 and NEP) by using different reading frames from the same RNA segment.

PA Encodes an RNA Polymerase

PB1 encodes an RNA polymerase and PB1-F2 protein (induces apoptosis) by using different reading frames from the same RNA segment.

PB2 Encodes an RNA Polymerase

The genome segments have common terminal sequences, and the ends of the RNA strands are partially complementary, allowing them to bond to each other by hydrogen bonds. After transcription from negative-sense to positive-sense RNA the +RNA strands get the cellular 5' cap added by cap snatching, which involves the viral protein NS1 binding to the cellular pre-mRNAs. The cap is then cleaved from the cellular pre-mRNA using a second viral protein, PA. The short oligo cap is then added to the influenza +RNA strands, allowing its processing as messenger RNA by ribosomes. The +RNA strands also serve for synthesis of -RNA strands for new virions. The RNA synthesis and its assembly with the nucleoprotein takes place in the cell nucleus, the synthesis of proteins takes place in the cytoplasm. The assembled virion cores leave the nucleus and migrate towards the cell membrane, with patches of viral transmembrane proteins (hemagglutinin, neuraminidase and M2 proteins) and an underlying layer of the M1 protein, and bud through these patches, releasing finished enveloped viruses into the extracellular fluid.

Evidence has been found for the presence of RNA structure in Influenza A virus. Structured RNA appears to be favoured in the +RNA, and is clustered near regions with functional importance. Two pseudoknot RNA structures were predicted that include the 3' splice site in Segments 7 and 8, which are both alternatively spliced to produce the M2 and NEP proteins respectively. Towards the 5' splice sites of these mRNAs, hairpin loop and multibranch loop structures were predicted. As well as these structures near splice sites, a pseudoknot was proposed in Segment 2, which encompasses the start codon for an internally transcribed open reading frame for a small polypeptide product (PB1-F2).

In Nonhumans

Fowl act as natural asymptomatic carriers of Influenza A viruses. Prior to the current H5N1 epizootic, strains of Influenza A virus had been demonstrated to be transmitted from wild fowl to only birds, pigs, horses, seals, whales and humans; and only between humans and pigs and between humans and domestic fowl; and not other pathways such as domestic fowl to horse. Wild aquatic birds are the natural hosts for a large variety of influenza A viruses. Occasionally viruses are transmitted from these birds to other species and may then cause devastating outbreaks in domestic poultry or give rise to human influenza pandemics.

H5N1 has been shown to be transmitted to tigers, leopards, and domestic cats that were fed uncooked domestic fowl (chickens) with the virus. H3N8 viruses from horses have crossed over and caused outbreaks in dogs. Laboratory mice have been infected successfully with a variety of avian flu genotypes. Influenza A viruses spread in the air and in manure and survives longer in cold weather. It can also be transmitted by contaminated feed, water, equipment and clothing; however, there is no evidence that the virus can survive in well-cooked meat. Symptoms in animals vary, but virulent strains can cause death within a few days.

> *"Highly pathogenic avian influenza virus is on every top ten list available for potential agricultural bioweapon agents".*

Avian influenza viruses that the OIE and others test for in order to control poultry disease include: H5N1, H7N2, H1N7, H7N3, H13N6, H5N9, H11N6, H3N8, H9N2, H5N2, H4N8, H10N7, H2N2, H8N4, H14N5, H6N5, H12N5 and others.

***Table** : Known outbreaks of highly pathogenic flu in poultry 1959-2003*

Year	***Area***	***Affected***	***Subtype***
1959	Scotland	Chicken	H5N1
1963	England	Turkey	H7N3
1966	Ontario (Canada) Turkey	H5N9	
1976	Victoria (Australia) Chicken	H7N7	
1979	Germany	Chicken	H7N7
1979	England	Turkey	H7N7
1983	Pennsylvania (USA)* chicken, turkey	H5N2	
1983	Ireland	Turkey	H5N8
1985	Victoria (Australia) Chicken	H7N7	
1991	England	Turkey	H5N1
1992	Victoria (Australia) Chicken	H7N3	

Contd...

Year	***Area***	***Affected***	***Subtype***
1994	Queensland (Australia) Chicken	H7N3	
1994	Mexico*	Chicken	H5N2
1994	Pakistan* Chicken	H7N3	
1997	New South Wales (Australia) Chicken	H7N4	
1997	Hong Kong (China)* Chicken	H5N1	
1997	Italy	Chicken	H5N2
1999	Italy*	Turkey	H7N1
2002	Hong Kong (China) Chicken	H5N1	
2002	Chile	Chicken	H7N3
2003	Netherlands* Chicken	H7N7	

Outbreaks with significant spread to numerous farms, resulting in great economic losses. Most other outbreaks involved little or no spread from the initially infected farms.

1979: "More than 400 harbour seals, most of them immature, died along the New England coast between December 1979 and October 1980 of acute pneumonia associated with influenza virus, A/Seal/Mass/1/180 (H7N7)."

1995: "Vaccinated birds can develop asymptomatic infections that allow virus to spread, mutate, and recombine (ProMED-mail, 2004). Intensive surveillance is required to detect these "silent epidemics" in time to curtail them. In Mexico, for example, mass vaccination of chickens against epidemic H5N2 influenza in 1995 has had to continue in order to control a persistent and evolving virus (Lee et al., 2004)."

1997: "Influenza A viruses normally seen in one species sometimes can cross over and cause illness in another species. For example, until 1997, only H1N1 viruses circulated widely in the U.S. pig population. However, in 1997, H3N2 viruses from humans were introduced into the pig population and caused widespread disease among pigs. Most recently, H3N8 viruses from horses have crossed over and caused outbreaks in dogs."

2000: "In California, poultry producers kept their knowledge of a recent H6N2 avian influenza outbreak to themselves due to their fear of public rejection of poultry products; meanwhile, the disease spread across the western United States and has since become endemic."

2003: In Netherlands H7N7 influenza virus infection broke out in poultry on several farms.

2004: In North America, the presence of avian influenza strain H7N3 was confirmed at several poultry farms in British Columbia in February 2004. As of April 2004, 18 farms had been quarantined to halt the spread of the virus.

2005: Tens of millions of birds died of H5N1 influenza and hundreds of millions of birds were culled to protect humans from H5N1. H5N1 is endemic in birds in southeast Asia and represents a long term pandemic threat.

2006: H5N1 spreads across the globe killing hundreds of millions of birds and over 100 people causing a significant H5N1 impact from both actual deaths and predicted possible deaths.

Swine Flu

Swine influenza (or "pig influenza") refers to a subset of Orthomyxoviridae that create influenza in pigs and are endemic in pigs. The species of Orthomyxoviridae that can cause flu in pigs are Influenza A virus and Influenza C virus but not all genotypes of these two species infect pigs. The known subtypes of Influenza A virus that create influenza in pigs and are endemic in pigs are H1N1, H1N2, H3N1 and H3N2.

Horse Flu

Horse flu (or "Equine influenza") refers to varieties of Influenza A virus that affect horses. Horse 'flu viruses were only isolated in 1956. There are two main types of virus called equine-1 (H7N7) which commonly affects horse heart muscle and equine-2 (H3N8) which is usually more severe.

Dog Flu

Dog flu (or "canine influenza") refers to varieties of Influenza A virus that affect dogs. The equine influenza virus H3N8 was found to infect and kill - with respiratory illness - greyhound race dogs at a Florida racetrack in January 2004.

H3N8

H3N8 is now endemic in birds, horses and dogs.

Human Influenza Virus

"Human influenza virus" usually refers to those subtypes that spread widely among humans. H1N1, H1N2, and H3N2 are the only known Influenza A virus subtypes currently circulating among humans.

Genetic factors in distinguishing between "human flu viruses" and "avian influenza viruses" include:

PB2: (RNA polymerase): Amino acid (or residue) position 627 in the PB2 protein encoded by the PB2 RNA gene. Until H5N1, all known avian influenza viruses had a Glu at position 627, while all human influenza viruses had a lysine.

HA: (hemagglutinin): Avian influenza HA bind alpha 2-3 sialic acid receptors while human influenza HA bind alpha 2-6 sialic acid receptors. Swine influenza viruses have the ability to bind both types of sialic acid receptors.

> *"About 52 key genetic changes distinguish avian influenza strains from those that spread easily among people, according to researchers in Taiwan, who analysed the genes of more than 400 A type flu viruses." "How many mutations would make an avian virus capable of infecting humans efficiently, or how many mutations would render an influenza virus a pandemic strain, is difficult to predict. We have examined sequences from the 1918 strain, which is the only pandemic influenza virus that could be entirely derived from avian strains. Of the 52 species-associated positions, 16 have residues typical for human strains; the others remained as avian signatures. The result supports the hypothesis that the 1918 pandemic virus is more closely related to the avian influenza A virus than are other human influenza viruses."*

Human flu symptoms usually include fever, cough, sore throat, muscle aches, conjunctivitis and, in severe cases, severe breathing problems and pneumonia that may be fatal.

The severity of the infection will depend to a large part on the state of the infected person's immune system and if the victim has been exposed to the strain before, and is therefore partially immune.

Highly pathogenic H5N1 avian influenza in a human is far worse, killing 50% of humans that catch it. In one case, a boy with H5N1 experienced diarrhea followed rapidly by a coma without developing respiratory or flu-like symptoms.

The Influenza A virus subtypes that have been confirmed in humans, ordered by the number of known human pandemic deaths, are:

H1N1 caused "Spanish Flu" and the 2009 swine flu outbreak

H2N2 caused "Asian Flu" in the late 1950s

H3N2 caused "Hong Kong Flu" in the late 1960s

H5N1 considered a global influenza pandemic threat through its spread in the mid 2000s

H7N7 has unusual zoonotic potential

H1N2 is currently endemic in humans and pigs

H9N2, H7N2, H7N3, H5N2, H10N7.

H1N1

H1N1 is currently pandemic in both human and pig populations. A variant of H1N1 was responsible for the Spanish flu pandemic that killed some 50 million to 100 million people worldwide over about a year in 1918 and 1919.

Another variant was named a pandemic threat in the 2009 flu pandemic. Controversy arose in October, 2005, after the H1N1 genome was published in the journal, *Science*, because of fears that this information could be used for bioterrorism.

H2N2

The Asian Flu was a pandemic outbreak of H2N2 avian influenza that originated in China in 1957, spread worldwide that same year during which a influenza vaccine was developed, lasted until 1958 and caused between one and four million deaths.

H3N2

H3N2 is currently endemic in both human and pig populations. It evolved from H2N2 by antigenic shift and caused the Hong Kong Flu pandemic of 1968 and 1969 that killed up to 750,000. "An early-onset, severe form of influenza A H3N2 made headlines when it claimed the lives of several children in the United States in late 2003."

The dominant strain of annual flu in January 2006 was H3N2. Measured resistance to the standard antiviral drugs amantadine and rimantadine in H3N2 increased from 1% in 1994 to 12% in 2003 to 91% in 2005.

> *"Contemporary human H3N2 influenza viruses are now endemic in pigs in southern China and can reassort with avian H5N1 viruses in this intermediate host."*

H5N1

H5N1 is the world's major influenza pandemic threat.

> *"When he compared the 1918 virus with today's human flu viruses, Dr. Taubenberger noticed that it had alterations in just 25 to 30 of the virus's 4,400 amino acids. Those few changes turned a bird virus into a killer that could spread from person to person."*

H7N7

H7N7 has unusual zoonotic potential. In 2003 in Netherlands 89 people were confirmed to have H7N7 influenza virus infection following an outbreak in poultry on several farms. One death was recorded.

H1N2

H1N2 is currently endemic in both human and pig populations. The new H1N2 strain appears to have resulted from the reassortment of the genes of the currently circulating influenza H1N1 and H3N2 subtypes.

The hemagglutinin protein of the H1N2 virus is similar to that of the currently circulating H1N1 viruses and the neuraminidase protein is similar to that of the current H3N2 viruses.

H9N2

Low pathogenic avian influenza A (H9N2) infection was confirmed in 1999, in China and Hong Kong in two children, and in 2003 in Hong Kong in one child. All three fully recovered.

H7N2

One person in New York in 2003 and one person in Virginia in 2002 were found to have serologic evidence of infection with H7N2. Both fully recovered.

H7N3

In North America, the presence of avian influenza strain H7N3 was confirmed at several poultry farms in British Columbia in February 2004.

As of April 2004, 18 farms had been quarantined to halt the spread of the virus. Two cases of humans with avian influenza have been confirmed in that region. "Symptoms included conjunctivitis and mild influenza-like illness." Both fully recovered.

H5N2

Japan's Health Ministry said January 2006 that poultry farm workers in Ibaraki prefecture may have been exposed to H5N2 in 2005. The H5N2 antibody titres of paired sera of 13 subjects increased fourfold or more.

H10N7

In 2004 in Egypt H10N7 was reported for the first time in humans. It caused illness in two infants in Egypt. One child's father is a poultry merchant.

Evolution

Taubenberger Says

"All influenza A pandemics since the Spanish flu pandemic, and indeed almost all cases of influenza A worldwide (excepting human infections from avian viruses such as H5N1 and H7N7), have been caused by descendants of the 1918 virus, including "drifted" H1N1 viruses and reassorted H2N2 and H3N2 viruses. The latter are composed of key genes from the 1918 virus, updated by subsequently incorporated avian influenza genes that code for novel surface proteins, making the 1918 virus indeed the "mother" of all pandemics.

Researchers from the National Institutes of Health used data from the Influenza Genome Sequencing Project and concluded that during the ten-year period examined most of the time the hemagglutinin gene in H3N2 showed no significant excess of mutations in the antigenic regions while an increasing variety of strains accumulated. This resulted in one of the variants eventually achieving higher fitness, becoming dominant, and in a brief interval of rapid evolution rapidly sweeping through the population and eliminating most other variants.

Avian influenza A(H5N1)- update 31: Situation (poultry) in Asia: need for a long-term response, comparison with previous outbreaks:

During last week's emergency meeting in Bangkok, Thailand: officials from FAO, OIE, and WHO drew attention to several unique features of the current outbreaks of H5N1 in poultry in Asia, in particular its geographical distribution, rate of spread and severity of which are unprecedented.

Prospects for rapid control are inconsistent with worldwide experience: over more than four decades, with previous outbreaks, which have all been much smaller in scope and inherently less challenging. Even in countries with good surveillance, adequate resources, and geographically limited outbreaks, control has often taken up to two years. For these reasons and others, WHO has cautioned against assumptions that the outbreaks can be controlled in the immediate future.

WHO has described the serious public health implications of these outbreaks in a previous update.

Up to the end of 2003: highly pathogenic avian influenza (HPAI) was considered a rare disease. Since 1959, only 21 Outbreakshad been reported worldwide. The majority occurred in Europe and the Americas. Of the total, only five resulted in significant spread to numerous farms, and only one was associated with spread to other countries.

Since mid-December 2003: eight Asian countries have confirmed outbreaks of highly pathogenic avian influenza caused by the H5N1 strain. Most of these countries are experiencing outbreaks of this disease for the first time in their histories. In several, outbreaks have been detected in virtually every part of the country

Over the past two months: more than 100 million birds have either died of the disease or been culled in Asia. This figure is greater than the total number of poultry affected, over years, in the world's previous five largest outbreaks combined.

Worldwide experience since 1959 supports official statements about the unprecedented nature of the present situation and the challenges for control. Unique features in the present situation include:

Concentration of poultry in backyard farms: In several countries experiencing outbreaks, up to 80% of poultry are produced on small farms and backyard holdings in rural areas, where poultry range freely. In China, 60% of the country's esitmated 13.2 billion chickens are raised on small farms in close proximity to humans and domestic animals, including pigs. This situation makes implementation of strict control measures, essential to the control of previous outbreaks, extremely difficult. These control measures – including bird-proof, ecologically controlled housing, treatment of water supplies, disinfection of all incoming persons, equipment, and vehicles, prevention of contact with insects, rodents, and other mechanical vectors – cannot be applied on small rural farms and backyard holdings.

Economic significance of poultry production: Poultry production contributes greatly to the economies and food supplies of affected countries. The agricultural sector faces the challenge of minimising losses to industry and subsistence farmers in ways that also reduce health risks for humans. Because many people in the region are so dependent on poultry, appropriate culling may be difficult to implement.

Lack of control experience: Since the disease is new to most countries in the region, very little experience exists at national and international levels to guide the best country-specific control measures.

In some countries, announcements of successful culling in certain areas are being followed by subsequent eruptions of disease in the same areas, suggesting reintroduction of the virus, continuing presence in the environment, or inadequate verification of outbreak control.

Lack of resources: Several countries with very widespread outbreaks lack adequate infrastructure and resources, including resources to compensate farmers and thus encourage compliance with government recommendations.

In some countries that have announced outbreaks, neither surveillance to detect the extent of spread nor culling of animals known to be infected is taking place.

The scale of international spread: With so many adjacent countries affected, a region-wide strategy will be needed to ensure that gains in one country are not compromised by inadequate control in another.

These unique features will make rapid control and long-term prevention of recurrence extremely difficult to achieve.

Culling remains the first line of action, as recommended by FAO, OIE, and WHO, for bringing the current outbreaks under control. Unlike other economically important domestic animals, poultry raising takes place in a very short production system. Provided sufficient resources are available to replace culled poultry stock, countries should not postpone aggressive culling because of fears of long-term consequences on poultry production.

Wild birds can play a role in introducing a virus of low pathogenicity into domestic flocks where, if allowed to circulate for several months, it can mutate into a highly pathogenic form. No evidence to date indicates that wild birds are the source of the present outbreaks of highly pathogenic H5N1 avian influenza. Wild birds should not be culled.

***Table** : Previous outbreaks of highly pathogenic avian influenza worldwide*

Year	*Country/area*	*Domestic birds affected*	*Strain*
1959	Scotland	chicken	H5N1
1963	England	turkey	H7N3
1966	Ontario (Canada)	turkey	H5N9
1976	Victoria (Australia)	chicken	H7N7
1979	Germany	chicken	H7N7
1979	England	turkey	H7N7
1983–1985	Pennsylvania (USA)* chicken,	turkey	H5N2
1983	Ireland	turkey	H5N8
1985	Victoria (Australia)	chicken	H7N7
1991	England	turkey	H5N1
1992	Victoria (Australia)	chicken	H7N3
1994	Queensland (Australia)	chicken	H7N3
1994–1995	Mexico*	chicken	H5N2
1994	Pakistan*	chicken	H7N3
1997	New South Wales (Australia)	chicken	H7N4
1997	Hong Kong (China)*	chicken	H5N1
1997	Italy	chicken	H5N2
1999–2000	Italy*	turkey	H7N1
2002	Hong Kong (China)	chicken	H5N1
2002	Chile	chicken	H7N3
2003	Netherlands*	chicken	H7N7

Outbreaks with significant spread to numerous farms, resulting in great economic losses. Most other outbreaks involved little or no spread from the initially infected farms.

Observations from Previous Outbreaks (1959–2003)

Outbreaks of highly pathogenic avian influenza can be extremely difficult to control, even under favourable conditions (concentration of infected birds in well-maintained commercial production facilities, limited geographical occurrence).

The 1983 Pennsylvania (USA) outbreak took two years to control. Some 17 million birds were destroyed at a direct cost of US$62 million. Indirect costs have been estimated at more than US$250 million. The 2003 outbreak in the Netherlands spread to Belgium and Germany.

In the Netherlands, more than 30 million birds - a quarter of the country's poultry stock – were destroyed. Some 2.7 million were destroyed in Belgium, and around 400,000 in Germany. In the Netherlands, 89 humans were infected, of whom one (a veterinarian) died. In that outbreak, measures needed to protect the health of poultry workers, farmers, and persons visiting farms included wearing of protective clothing, masks to cover the mouth and nose, eye protection, vaccination against normal seasonal human influenza, and administration of prophylactic antiviral drugs.

Control is Even more Difficult in Countries with Dense Poultry Populations

The Italian outbreak of 1999–2000 caused infection in 413 flocks, including 25 backyard flocks, and resulted in the destruction of around 14 million birds. Control was complicated by the occurrence of cases in areas with extremely dense poultry populations.

Compensation to farmers amounted to US$63 million. Costs for the poultry and associated industry have been estimated at US$620 million. Four months after the last outbreak ended, the virus returned in a low-pathogenic form, rapidly causing a further 52 outbreaks.

Although the last outbreak of highly pathogenic avian influenza in Mexico occurred in 1995, the causative agent – the H5N2 strain – has never been entirely eliminated from the country, in its present low-pathogenicity form, despite years of intense efforts, including the administration of more than 2 billion doses of vaccines of varying efficacy. Similarly, the vaccination policy pursued in Pakistan does not appear to have resulted in eradication of the causative agent.

Avoidance of contact between poultry and wild birds, especially ducks and other waterfowl, can help prevent the introduction of a low-pathogenicity virus into domestic flocks. Although no evidence to date has conclusively linked the current outbreaks with wild migratory birds in Asia:

Several of these outbreaks have been linked to contact between free-ranging flocks and wild birds, including the shared use of water sources. Faecal contamination of water supplies is considered a very efficient way for waterfowl to transmit the virus. Virus (low-pathogenicity) has been readily recovered from lakes and ponds where migratory birds congregate. An especially risky practice is the raising of small numbers of domestic ducks on a pond in proximity to domestic chicken and turkey flocks. Domestic ducks attract wild ducks, and provide a significant link in the chain of transmission from wild birds to domestic flocks. Aggressive control measures, including culling of infected and exposed poultry, are recommended for avian influenza virus subtypes H5 and H7 even when the virus initially shows low pathogenicity. (H5 and H7 are the only subtypes implicated in outbreaks of highly pathogenic disease.)

Several of the largest outbreaks (Pennsylvania, Mexico, Italy) initially began with mild illness in poultry.

When the virus was allowed to continue circulating in poultry, it eventually mutated (within 6 to 9 months) into a highly pathogenic form with a mortality ratio approaching 100%. Moreover, the initial presence of low-pathogenicity virus in these outbreaks complicated diagnosis of the highly pathogenic form.

Avian Influenza A (H5N1) Infection in Humans

An unprecedented epizootic avian influenza A (H5N1) virus that is highly pathogenic has crossed the species barrier in Asia to cause many human fatalities and poses an increasing pandemic threat.

This summary describes the features of human infection with influenza A (H5N1) and reviews recommendations for prevention and clinical management presented in part at the recent World Health Organisation (WHO) Meeting on Case Management and Research on Human Influenza A/H5, which was held in Hanoi, May 10 through 12, 2005.1 Because many critical questions remain, modifications of these recommendations are likely.

Incidence

The occurrence of human influenza A (H5N1) in Southeast Asia has paralleled large outbreaks of avian influenza A (H5N1), although the avian epidemics in 2004 and 2005 have only rarely led to disease in humans.

The largest number of cases has occurred in Vietnam, particularly during the third, ongoing wave, and the first human death was recently reported in Indonesia.

The frequencies of human infection have not been determined, and seroprevalence studies are urgently needed. The expanding geographic distribution of avian influenza A (H5N1) infections, with recent outbreaks in Kazakstan, Mongolia, and Russia, indicates that more human populations are at risk.

Transmission

Human influenza is transmitted by inhalation of infectious droplets and droplet nuclei, by direct contact, and perhaps, by indirect (fomite) contact, with self-inoculation onto the upper respiratory tract or conjunctival mucosa.4, 5 The relative efficiency of the different routes of transmission has not been defined. For human influenza A (H5N1) infections, evidence is consistent with bird-to-human, possibly environment-to-human, and limited, nonsustained human-to-human transmission to date.

Animal to Human

In 1997, exposure to live poultry within a week before the onset of illness was associated with disease in humans, whereas there was no significant risk related to eating or preparing poultry products or exposure to persons with influenza A (H5N1) disease. Exposure to ill poultry and butchering of birds were associated with seropositivity for influenza A (H5N1). Recently, most patients have had a history of direct contact with poultry, although not those who were involved in mass culling of poultry. Plucking and preparing of diseased birds; handling fighting cocks; playing with poultry, particularly asymptomatic infected ducks; and consumption of duck's blood or possibly undercooked poultry have all been implicated. Transmission to felids has been observed by feeding raw infected chickens to tigers and leopards in zoos in Thailand and to domestic cats under experimental conditions. Transmission between felids has been found under such conditions. Some infections may be initiated by pharyngeal or gastrointestinal inoculation of virus.

Human to Human

Human-to-human transmission of influenza A (H5N1) has been suggested in several household clusters and in one case of apparent child-to-mother transmission. Intimate contact without the use of precautions was implicated, and so far no case of human-to-human transmission by small-particle aerosols has been identified. In 1997, human-to-human transmission did not apparently occur through social contact, and serologic studies of exposed health care workers indicated that transmission was inefficient. Serologic surveys in Vietnam and Thailand have not found evidence of asymptomatic infections among contacts. Recently, intensified surveillance of contacts of patients by reverse-transcriptase–polymerase-chain-reaction (RT-PCR) assay has led to the detection of mild cases, more infections in older adults, and an increased number and duration of clusters in families in northern Vietnam, findings suggesting that the local virus strains may be adapting to humans. However, epidemiologic and virologic studies are needed to confirm these findings. To date, the risk of nosocomial transmission to health care workers has been low, even when appropriate isolation measures were not used. However, one case of severe illness was reported in a nurse exposed to an infected patient in Vietnam.

Environment to Human

Given the survival of influenza A (H5N1) in the environment, several other modes of transmission are theoretically possible. Oral ingestion of contaminated water during swimming and direct intranasal or conjunctival inoculation during exposure to water are other potential modes, as is contamination of hands from infected fomites and subsequent self-inoculation. The widespread use of untreated poultry feces as fertilizer is another possible risk factor.

Clinical Features

The clinical spectrum of influenza A (H5N1) in humans is based on descriptions of hospitalised patients. The frequencies of milder illnesses, subclinical infections, and atypical presentations (e.g., encephalopathy and gastroenteritis) have not been determined, but case reports indicate that each occurs. Most patients have been previously healthy young children or adults.

Incubation

The incubation period of avian influenza A (H5N1) may be longer than for other known human influenzas. In 1997, most cases occurred within two to four days after exposure; recent reports, indicate similar intervals but with ranges of up to eight days. The case-to-case intervals in household clusters have generally been 2 to 5 days, but the upper limit has been 8 to 17 days, possibly owing to unrecognised exposure to infected animals or environmental sources.

Initial Symptoms

Most patients have initial symptoms of high fever (typically a temperature of more than 38°C) and an influenza-like illness with lower respiratory tract symptoms. Upper respiratory tract symptoms are present only sometimes. Unlike patients with infections caused by avian influenza A (H7) viruses, patients with avian influenza A (H5N1) rarely have conjunctivitis. Diarrhea, vomiting, abdominal pain, pleuritic pain, and bleeding from the nose and gums have also been reported early in the course of illness in some patients. Watery diarrhea without blood or inflammatory changes appears to be more common than in influenza due to human viruses and may precede respiratory manifestations by up to one week. One report described two patients who presented with an encephalopathic illness and diarrhea without apparent respiratory symptoms.

Clinical Course

Lower respiratory tract manifestations develop early in the course of illness and are usually found at presentation. In one series, dyspnea developed a median of 5 days after the onset of illness (range, 1 to 16). Respiratory distress, tachypnea, and inspiratory crackles are common. Sputum production is variable and sometimes bloody. Almost all patients have clinically apparent pneumonia; radiographic changes include diffuse, multifocal, or patchy infiltrates; interstitial infiltrates; and segmental or lobular consolidation with air bronchograms. Radiographic abnormalities were present a median of 7 days after the onset of fever in one study (range, 3 to 17).15 In Ho Chi Minh City, Vietnam, multifocal consolidation involving at least two zones was the most common abnormality among patients at the time of admission. Pleural effusions are uncommon. Limited microbiologic data indicate that this process is a primary viral pneumonia, usually without bacterial suprainfection at the time of hospitalisation.

Progression to respiratory failure has been associated with diffuse, bilateral, ground-glass infiltrates and manifestations of the acute respiratory distress syndrome (ARDS). In Thailand, 15 the median time from the onset of illness to ARDS was 6 days (range, 4 to 13). Multiorgan failure with signs of renal dysfunction and sometimes cardiac compromise, including cardiac dilatation and supraventricular tachyarrhythmias, has been common. Other complications have included ventilator-associated pneumonia, pulmonary hemorrhage, pneumothorax, pancytopenia, Reye's syndrome, and sepsis syndrome without documented bacteremia.

Mortality

The fatality rate among hospitalised patients has been high, although the overall rate is probably much lower. In contrast to 1997, when most deaths occurred among patients older than 13 years of age, recent avian influenza A (H5N1) infections have caused high rates of death among infants and young children. The case fatality rate was

89 percent among those younger than 15 years of age in Thailand. Death has occurred an average of 9 or 10 days after the onset of illness (range, 6 to 30),15,16 and most patients have died of progressive respiratory failure.

Laboratory Findings

Common laboratory findings have been leukopenia, particularly lymphopenia; mild-to-moderate thrombocytopenia; and slightly or moderately elevated aminotransferase levels. Marked hyperglycemia, perhaps related to corticosteroid use, and elevated creatinine levels also occur. In Thailand, an increased risk of death was associated with decreased leukocyte, platelet, and particularly, lymphocyte counts at the time of admission.

Virologic Diagnosis

Antemortem diagnosis of influenza A (H5N1) has been confirmed by viral isolation, the detection of H5-specific RNA, or both methods. Unlike human influenza A infection, 26 avian influenza A (H5N1) infection may be associated with a higher frequency of virus detection and higher viral RNA levels in pharyngeal than in nasal samples.

In Vietnam, the interval from the onset of illness to the detection of viral RNA in throat-swab samples ranged from 2 to 15 days (median, 5.5), and the viral loads in pharyngeal swabs 4 to 8 days after the onset of illness were at least 10 times as high among patients with influenza A (H5N1) as among those with influenza A (H3N2) or (H1N1). Earlier studies in Hong Kong also found low viral loads in nasopharyngeal samples. Commercial rapid antigen tests are less sensitive in detecting influenza A (H5N1) infections than are RT-PCR assays. In Thailand, the results of rapid antigen testing were positive in only 4 of 11 patients with culture-positive influenza A (H5N1) (36 percent) 4 to 18 days after the onset of illness.

Management

Most hospitalised patients with avian influenza A (H5N1) have required ventilatory support within 48 hours after admission, as well as intensive care for multiorgan failure and sometimes hypotension. In addition to empirical treatment with broad-spectrum antibiotics, antiviral agents, alone or with corticosteroids, have been used in most patients, although their effects have not been rigorously assessed. The institution of these interventions late in the course of the disease has not been associated with an apparent decrease in the overall mortality rate, although early initiation of antiviral agents appears to be beneficial. Cultivable virus generally disappears within two or three days after the initiation of oseltamivir among survivors, but clinical progression despite early therapy with oseltamivir and a lack of reductions in pharyngeal viral load have been described in patients who have died.

Pathogenesis

Characterisation of Virus

Studies of isolates of avian influenza A (H5N1) from patients in 1997 revealed that virulence factors included the highly cleavable hemagglutinin that can be activated by multiple cellular proteases, a specific substitution in the polymerase basic protein 2 (Glu627Lys) that enhances replication, and a substitution in nonstructural protein 1 (Asp92Glu) that confers increased resistance to inhibition by interferons and tumour necrosis factor á (TNF-á) in vitro and prolonged replication in swine, as well as greater elaboration of cytokines, particularly TNF-á, in human macrophages exposed to the virus. Since 1997, studies of influenza A (H5N1) indicate that these viruses continue to evolve, with changes in antigenicity and internal gene constellations; an expanded host range in avian species, and the ability to infect fields; enhanced pathogenicity in experimentally infected mice and ferrets, in which they cause systemic infections; and increased environmental stability.

Phylogenetic analyses indicate that the Z genotype has become dominant and that the virus has evolved into two distinct clades, one encompassing isolates from Cambodia, Laos, Malaysia, Thailand, and Vietnam and the other isolates from China, Indonesia, Japan, and South Korea. Recently, a separate cluster of isolates has appeared in northern Vietnam and Thailand, which includes variable changes near the receptor-binding site and one fewer arginine residue in the polybasic cleavage site of the hemagglutinin. However, the importance of these genetic and biologic changes with respect to human epidemiology or virulence is uncertain.

Patterns of Viral Replication

The virologic course of human influenza A (H5N1) is incompletely characterised, but studies of hospitalised patients indicate that viral replication is prolonged. In 1997, virus could be detected in nasopharyngeal isolates for a median of 6.5 days (range, 1 to 16), and in Thailand, the interval from the onset of illness to the first positive culture ranged from 3 to 16 days. Nasopharyngeal replication is less than in human influenza, and studies of lower respiratory tract replication are needed. The majority of faecal samples tested have been positive for viral RNA (seven of nine), whereas urine samples were negative. The high frequency of diarrhea among affected patients and the detection of viral RNA in faecal samples, including infectious virus in one case, suggest that the virus replicates in the gastrointestinal tract. The findings in one autopsy confirmed this observation.

Highly pathogenic influenza A (H5N1) viruses possess the polybasic amino acid sequence at the hemagglutinin-cleavage site that is associated with visceral dissemination in avian species. Invasive infection has been documented in mammals, and in humans, six of six serum specimens were positive for viral RNA four to nine days after the onset of illness. Infectious virus and RNA were detected in blood, cerebrospinal fluid, and feces in one patient. Whether feces or blood serves to transmit infection under some circumstances is unknown.

Host Immune Responses

The relatively low frequencies of influenza A (H5N1) illness in humans despite widespread exposure to infected poultry indicate that the species barrier to acquisition of this avian virus is substantial. Clusters of cases in family members may be caused by common exposures, although the genetic factors that may affect a host's susceptibility to disease warrant study.

The innate immune responses to influenza A (H5N1) may contribute to disease pathogenesis. In the 1997 outbreaks, elevated blood levels of interleukin-6, TNF-á, interferon-ã, and soluble interleukin-2. receptor were observed in individual patients, and in the patients in 2003, elevated levels of the chemokines interferon-inducible protein 10, monocyte chemoattractant protein 1, and monokine induced by interferon-ã were found three to eight days after the onset of illness.

Recently, plasma levels of inflammatory mediators (interleukin-6, interleukin-8, interleukin-1â, and monocyte chemoattractant protein 1) were found to be higher among patients who died than among those who survived (Simmons C: personal communication), and the average levels of plasma interferon-á were about three times as high among patients with avian influenza A who died as among healthy controls. Such responses may be responsible in part for the sepsis syndrome, ARDS, and multiorgan failure observed in many patients.

Among survivors, specific humoral immune responses to influenza A (H5N1) are detectable by microneutralisation assay 10 to 14 days after the onset of illness. Corticosteroid use may delay or blunt these responses.

Pathological Findings

Limited postmortem analyses have documented severe pulmonary injury with histopathological changes of diffuse alveolar damage, consistent with findings in other reports of pneumonia due to human influenza virus. Changes include filling of the alveolar spaces with fibrinous exudates and red cells, hyaline-membrane formation, vascular congestion, infiltration of lymphocytes into the interstitial areas, and the proliferation of reactive fibroblasts.

Infection of type II pneumocytes occurs. Antemortem biopsy of bone marrow specimens has shown reactive histiocytosis with hemophagocytosis in several patients, and lymphoid depletion and atypical lymphocytes have been noted in spleen and lymphoid tissues at autopsy. Centrilobular hepatic necrosis and acute tubular necrosis have been noted in several instances.

Case Detection and Management

The possibility of influenza A (H5N1) should be considered in all patients with severe acute respiratory illness in countries or territories with animal influenza A (H5N1), particularly in patients who have been exposed to poultry. However, some outbreaks in poultry were recognised only after sentinel cases occurred in humans. Early recognition of cases is confounded by the nonspecificity of the initial clinical manifestations and high background rates of acute respiratory illnesses from other causes. In addition, the possibility of influenza A (H5N1) warrants consideration in patients presenting with serious unexplained illness (e.g., encephalopathy or diarrhea) in areas with known influenza A (H5N1) activity in humans or animals.

The diagnostic yield of different types of samples and virologic assays is not well defined. In contrast to infections with human influenza virus, throat samples may have better yields than nasal samples. Rapid antigen assays may help provide support for a diagnosis of influenza A infection, but they have poor negative predictive value and lack specificity for influenza A (H5N1). The detection of viral RNA in respiratory samples appears to offer the greatest sensitivity for early identification, but the sensitivity depends heavily on the primers and assay method used. Laboratory confirmation of influenza A (H5N1) requires one or more of the following: a positive viral culture, a positive PCR assay for influenza A (H5N1) RNA, a positive immunofluorescence test for antigen with the use of monoclonal antibody against H5, and at least a fourfold rise in H5-specific antibody titre in paired serum samples.

Hospitalisation

Whenever feasible while the numbers of affected persons are small, patients with suspected or proven influenza A (H5N1) should be hospitalised in isolation for clinical monitoring, appropriate diagnostic testing, and antiviral therapy. If patients are discharged early, both the patients and their families require education on personal hygiene and infection-control measures. Supportive care with provision of supplemental oxygen and ventilatory support is the foundation of management.1 Nebulisers and high–air flow oxygen masks have been implicated in the nosocomial spread of severe acute respiratory syndrome (SARS) and should be used only with strict airborne precautions.

Antiviral Agents

Patients with suspected influenza A (H5N1) should promptly receive a neuraminidase inhibitor pending the results of diagnostic laboratory testing. The optimal dose and duration of treatment with neuraminidase inhibitors are uncertain, and currently approved regimens likely represent the minimum required. These viruses are susceptible in vitro to oseltamivir and zanamivir. Oral oseltamivir and topical zanamivir are active in animal models of influenza A (H5N1). Recent murine studies indicate that as compared with an influenza A (H5N1) strain from 1997, the strain isolated in 2004 requires higher oseltamivir doses and more prolonged administration (eight days) to induce similar antiviral effects and survival rates. Inhaled zanamivir has not been studied in cases of influenza A (H5N1) in humans.

Early treatment will provide the greatest clinical benefit, although the use of therapy is reasonable when there is a likelihood of ongoing viral replication. Placebo-controlled clinical studies of oral oseltamivir and inhaled zanamivir comparing currently approved doses with doses that are twice as high found that the two doses had similar tolerability but no consistent difference in clinical or antiviral benefits in adults with uncomplicated human influenza. Although approved doses of oseltamivir (75 mg twice daily for five days in adults and weight-adjusted twice-daily doses for five days in children older than one year of age — twice-daily doses of 30 mg for those weighing 15 kg or less, 45 mg for those weighing more than 15 to 23 kg, 60 mg for those weighing more than 23 to 40 kg, and 75 mg for those weighing more than 40 kg) are reasonable for treating early, mild cases of influenza A (H5N1), higher doses (150 mg twice daily in adults) and treatment for 7 to 10 days are considerations in treating severe infections, but prospective studies are needed.

High-level antiviral resistance to oseltamivir results from the substitution of a single amino acid in N1 neuraminidase (His274Tyr). Such variants have been detected in up to 16 percent of children with human influenza A (H1N1) who have received oseltamivir.

Not surprisingly, this resistant variant has been detected recently in several patients with influenza A (H5N1) who were treated with oseltamivir.

Although less infectious in cell culture and in animals than susceptible parental virus, oseltamivir-resistant H1N1 variants are transmissible in ferrets. Such variants retain full susceptibility to zanamivir and partial susceptibility to the investigational neuraminidase inhibitor peramivir in vitro.

In contrast to isolates from the 1997 outbreak, recent human influenza A (H5N1) isolates are highly resistant to the M2 inhibitors amantadine and rimantadine, and consequently, these drugs do not have a therapeutic role. Agents of clinical investigational interest for treatment include zanamivir, peramivir, long-acting topical neuraminidase inhibitors, ribavirin and possibly, interferon alfa.

Immunomodulators

Corticosteroids have been used frequently in treating patients with influenza A (H5N1), with uncertain effects. Among five patients given corticosteroids in 1997, two treated later in their course for the fibroproliferative phase of ARDS survived. In a randomised trial in Vietnam, all four patients given dexamethasone died. Interferon alfa possesses both antiviral and immunomodulatory activities, but appropriately controlled trials of immunomodulatory interventions are needed before routine use is recommended.

Prevention

Immunisation

No influenza A (H5) vaccines are currently commercially available for humans. Earlier H5 vaccines were poorly immunogenic and required two doses of high hemagglutinin antigen content or the addition of MF adjuvant to generate neutralising antibody responses. A third injection of adjuvanted 1997 H5 vaccine variably induced cross-reacting antibodies to human isolates from 2004. Reverse genetics has been used for the rapid generation of nonvirulent vaccine viruses from recent influenza A (H5) isolates, and several candidate vaccines are under study.

One such inactivated vaccine with the use of a human H5N1 isolate from 2004 has been reported to be immunogenic at high hemagglutinin doses. Studies with approved adjuvants like alum are urgently needed. Live attenuated, cold-adapted intranasal vaccines are also under development. These are protective against human influenza after a single dose in young children.

Hospital-Infection Control

Influenza is a well-recognised nosocomial pathogen. Current recommendations are based on efforts to reduce transmission to health care workers and other patients in a nonpandemic situation and on the interventions used to contain SARS. The efficiency of surgical masks, even multiple ones, is much less than that of N-95 masks, but they could be used if the latter are not available. Chemoprophylaxis with 75 mg of oseltamivir once daily for 7 to 10 days is warranted for persons who have had a possible unprotected exposure. The use of preexposure prophylaxis warrants consideration if evidence indicates that the influenza A (H5N1) strain is being transmitted from person to person with increased efficiency or if there is a likelihood of a high-risk exposure (e.g., an aerosol-generating procedure).

Household and Close Contacts

Household contacts of persons with confirmed cases of influenza A (H5N1) should receive postexposure prophylaxis as described above. Contacts of a patient with proven or suspected virus should monitor their temperature and symptoms.

Although the risk of secondary transmission has appeared low to date, self-quarantine for a period of one week after the last exposure to an infected person is appropriate. If evidence indicates that person-to-person transmission may be occurring, quarantine of exposed contacts should be enforced. For others who have had an unprotected exposure to an infected person or to an environmental source (e.g., exposure to poultry) implicated in the transmission of influenza A (H5N1), postexposure chemoprophylaxis as described above may be warranted.

Conclusions

Infected birds have been the primary source of influenza A (H5N1) infections in humans in Asia. Transmission between humans is very limited at present, but continued monitoring is required to identify any increase in viral adaptation to human hosts. Avian influenza A (H5N1) in humans differs in multiple ways from influenza due to human viruses, including the routes of transmission, clinical severity, pathogenesis, and perhaps, response to treatment. Case detection is confounded by the nonspecificity of initial manifestations of illness, so that detailed contact and travel histories and knowledge of viral activity in poultry are essential. Commercial rapid antigen tests are insensitive, and confirmatory diagnosis requires sophisticated laboratory support. Unlike human influenza, avian influenza A (H5N1) may have higher viral titres in the throat than in the nose, and hence, analysis of throat swabs or lower respiratory samples may offer more sensitive means of diagnosis. Recent human isolates are fully resistant to M2 inhibitors, and increased doses of oral oseltamivir may be warranted for the treatment of severe illness. Despite recent progress, knowledge of the epidemiology, natural history, and management of influenza A (H5N1) disease in humans is incomplete. There is an urgent need for more coordination in clinical and epidemiologic research among institutions in countries with cases of influenza A (H5N1) and internationally.

The views expressed in this article do not necessarily reflect those of the WHO or other meeting sponsors.

We are indebted to the National Institute of Allergy and Infectious Diseases and the Wellcome Trust for their collaborative support of the WHO meeting; to Drs. Klaus Stohr and Alice Croisier of the Global Influenza Program at the WHO, Geneva; and to Drs. Peter Horby and Monica Guardo and the staff of the WHO Country Offices, Vietnam, for organising the WHO consultation and for support in the preparation of the manuscript; and to Diane Ramm for help in the preparation of the manuscript.

6

Strategies for Protecting Poultry from Coccidia

If ever organisms lived up to the label "parasite," it is those belonging to the order Coccidia. Not only do the single-cell protozoans of the *Eimeria* genus infest the nation's poultry flocks, costing American producers an estimated $600 million-plus annually in medication costs and lost production; they also invade and take shelter in the very cells marshaled by the chicken's immune system to defeat them.

Though a vaccine is available in this country against coccidiosis, its key ingredient is a low dose of the live parasite, which stimulates protective immunity. But Hyun S. Lillehoj, who is an immunologist with the ARS Immunology and Disease Resistance Laboratory at Beltsville, Maryland, says the presence of the live parasite may pose problems.

"The live parasite can cause disease in the bird," Lillehoj contends. "If the bird's immune system isn't functioning properly for some reason, the live parasite in the vaccine can overcome the immune system. Also, there can be a negative interaction within the bird with feed contaminants such as mycotoxins or with other infections that might be present, such as *salmonella* or *campylobacter*."

Complicating the vaccine situation is the existence of seven different species of *Eimeria*.

"An effective vaccine needs to incorporate elements from all seven species," says Lillehoj. "The vaccine that has the live parasite uses many of the seven strains. But the incidence of variant species of *Eimeria* in the field is increasing, and the live coccidia vaccine cannot protect effectively against all of them. It's also very labour-intensive to produce the live parasites." Lillehoj favours a different approach. She and her research team including support scientist Marjorie B. Nichols and technician Melody B. Lowe have devised a two-pronged strategy to thwart coccidia.

"The chicken's immune system produces cytotoxic T-cells whose function it is to target and destroy infected cells," explains Lillehoj. "That's part of nature's protective immune mechanism against this parasite.

"But there is a phase of the coccidia life cycle when the parasites are called sporozoites. These actually get inside the cytotoxic cells, which then cannot kill them, but instead deliver the parasites to the part of the intestine called the crypt epithelium, where they exit to develop." Once nestled in crypt epithelial cells, the thriving coccidia wreak havoc in the intestinal lining and interfere with the chicken's ability to absorb nutrients from the feed it has eaten. Result: The bird doesn't gain weight and may die.

In 1993, Lillehoj and ARS immunologist James M. Trout observed coccidia's commandeering of cytotoxic cells firsthand when they used two fluorescent, colour-stained monoclonal antibodies to cling to and track movement of both parasites and cytotoxic cells inside chicken intestines.

Green-stained monoclonal antibodies allowed them to see where the coccidia went; red-stained antibodies pinpointed the presence of the cytotoxic cells. Overlapping red and green colours proved the coccidia invaded the very cells that were supposed to protect the chicken against them.

Part of Lillehoj's plan is to block the initial invasion of those crucial infection-fighting cytotoxic cells by the coccidia.

"The sporozoite has to bind to the cytotoxic cell to get inside it," she explains. "Once it binds, it makes a little dent on the cell. The parasite has a retractable structure called a conoid that makes this dent. Then enzymes from the parasite act on the cell to make an opening for the parasite to get in.

"We've developed and have applied for a patent on a chicken monoclonal antibody that identifies a protein that the sporozoite uses to cling to the cytotoxic cell. In laboratory tests, this antibody actually blocks the sporozoite invasion of the cytotoxic cell."

Lillehoj is working with ARS molecular biologists Mark C. Jenkins and Kang D. Choi on ways to use the protein recognised by the antibody as a potential vaccine. Also promising as potential weapons are cytokines, substances produced naturally by the bird's white blood cells.

We have shown in laboratory tests that some cytokines inhibit development of the parasite," says Lillehoj.

"They also enhance cytotoxic activity by turning precursor cytotoxic cells into active cytotoxic cells. Once activated, these cytotoxic cells kill parasite-infected host cells. Cytokines also activate white blood cells called macrophages to devour the parasites."

Starting in 1995, Lillehoj collaborated with ARS molecular biologist Dante S. Zarlenga and scientists at Korea's Seoul National University to clone the chicken gene that controls manufacture of a cytokine called gamma-interferon. The research team has produced genetically engineered chicken gamma-interferon and is testing its protective powers in live chickens. Early results look promising, Lillehoj says.

"If this works, a bird that's treated with the gamma-interferon might still get infected with coccidia, but it might not lose as much weight or get as bad a case of coccidiosis," she explains. "You wouldn't want to completely block the infection, anyway, because then you wouldn't stimulate the bird's immune system to provide natural immunity against future coccidia infections or other opportunistic pathogens." Gamma-interferon may prove useful in the battle against coccidia in other ways as well, Lillehoj adds.

"Antigens are proteins from the parasite that stimulate an immune response from an animal's immune system," she points out. "It's been shown in mammalian cells that when you add gamma-interferon to a weak antigen, you get a greater immune response than if you just vaccinate with the antigen alone. Plans are under way to use gamma-interferon as an adjuvant to enhance the action of the vaccine."

Mass production of gamma-interferon may be tricky, Lillehoj says. In lab tests, attempts to reproduce the substance by inserting the gene for its production into fast-multiplying *E. coli* bacteria fell short of the scientists' expectations.

"The protein was not very effective when expressed in *E. coli*," Lillehoj admits. "But once you have a gene that expresses the protein, you can raise it in a mammalian cell line." One possible solution to the protection dilemma might be to identify an antigen common to all strains of coccidia and use that as the basis for a new vaccine. "We know of one such segment, but we're not ready to test it yet as a vaccine," says Lillehoj. The more common game plan—waiting to clean up coccidiosis in flocks after it occurs—is rapidly becoming a risky proposition, according to Lillehoj.

"The major problem is that the parasite develops drug resistance very quickly," she notes. "The main emphasis for control has been on drugs, but the coccidia have developed resistance to all the drugs ever tested against them." The ARS research team's multifaceted efforts come down to one simple goal: to mimic nature.

"In the field, once birds have been exposed to coccidia, they develop immunity," says Lillehoj. "We've been trying to figure out how chickens get that immunity. Over the years, we've learned a lot about how the parasite invades cells and stimulates natural immunity." —By Sandy Miller Hays, ARS.

Hyun S. Lillehoj is at the USDA ARS Immunology and Disease Resistance Laboratory, Bldg. 1043, 10300 Baltimore Ave., Beltsville, MD 20705-2350; phone (301) 504-6170.

"Two Strategies for Protecting Poultry From Coccidia" was published in the October 1996 issue of *Agricultural Research* magazine.

Common Cold

The common cold (also known as nasopharyngitis, acute viral rhinopharyngitis, acute coryza, or a cold) (Latin: *rhinitis acuta catarrhalis*) is a viral infectious disease of the upper respiratory system, caused primarily by rhinoviruses and coronaviruses. Common symptoms include a cough, sore throat, runny nose, and fever. There is no cure for the common cold, but symptoms usually resolve in 7 to 10 days, with some symptoms possibly lasting for up to three weeks.

The common cold is the most frequent infectious disease in humans with the average adult contracting two to four infections a year and the average child contracting between 6 and 12. Collectively, colds, influenza, and other upper respiratory tract infections (URTI) with similar symptoms are included in the diagnosis of influenza-like illness.

Signs and Symptoms

Symptoms are cough, sore throat, runny nose, and nasal congestion; sometimes this may be accompanied by conjunctivitis (pink eye), muscle aches, fatigue, headaches, shivering, and loss of appetite.

Fever is often present thus creating a symptom picture which overlaps with influenza. The symptoms of influenza however are usually more severe.

Those suffering from colds often report a sensation of chilliness even though the cold is not generally accompanied by fever, and although chills are generally associated with fever, the sensation may not always be caused by actual fever. In one study, 60% of those suffering from a sore throat and upper respiratory tract infection reported headaches, often due to nasal congestion.

Progression

The viral replication begins 2 to 6 hours after initial contact. Symptoms usually begin 2 to 5 days after initial infection but occasionally occur in as little as 10 hours. Symptoms peak 2–3 days after symptom onset, whereas influenza symptom onset is constant and immediate. The symptoms usually resolve spontaneously in 7 to 10 days but some can last for up to three weeks. In children the cough lasts for more than 10 days in 35–40% of cases and continues for more than 25 days in 10%.

The first indication of an upper respiratory virus is often a sore or scratchy throat. Other common symptoms are runny nose, congestion, and sneezing. These are sometimes accompanied by muscle aches, fatigue, malaise, headache, weakness, or loss of appetite. Cough and fever generally indicate influenza rather than an upper respiratory virus with a positive predictive value of around 80%. Symptoms may be more severe in infants and young children, and in these cases it may include fever and hives. Upper respiratory viruses may also be more severe in smokers.

Infectious Period

Researchers have studied rhinovirus-caused colds more than other colds. Rhinovirus-caused colds are most infectious during the first three days of symptoms. They become much less infectious after those three days.

Cause

Viruses

The common cold is a viral infection of the upper respiratory tract. The most commonly implicated virus is a rhinovirus (30–50%), a type of picornavirus with 99 known serotypes. Others include:

coronavirus (10–15%), influenza (5–15%), human parainfluenza viruses, human respiratory syncytial virus, adenoviruses, enteroviruses, and metapneumovirus.

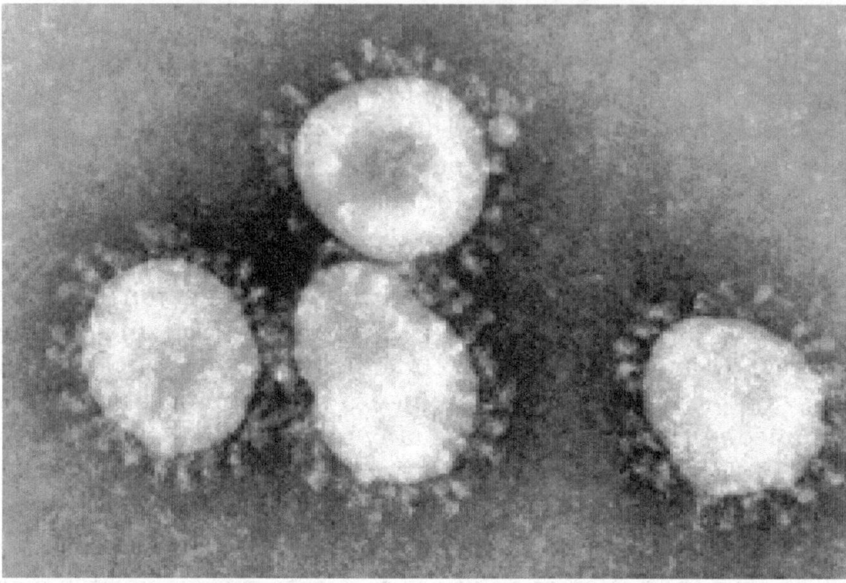

Figure : *Coronaviruses are a group of viruses known for causing the common cold. They have a halo, or crown-like (corona) appearance when viewed under an electron microscope.*

In total over 200 serologically different viral types cause colds. Coronaviruses are particularly implicated in adult colds. Of over 30 coronaviruses, 3 or 4 cause infections in humans, but they are difficult to grow in the laboratory and their significance is thus less well-understood. Due to the many different types of viruses and their tendency for continuous mutation, it is impossible to gain complete immunity to the common cold.

Risk Factors

Touching eyes, nose, or mouth with contaminated fingers. This behaviour increases the likelihood of transferring viruses from the surface of the hands, where they are harmless, into the upper respiratory tract, where they can infect the tissues. It has been demonstrated that cold viruses can be spread by touching contaminated objects and surfaces, or by brief contact of hands.

Spending time in an enclosed area with an infected person or in close contact with an infected person. Common colds are droplet-borne infections, which means that they can be transmitted through breathing in tiny particles that the infected person emits when he or she coughs or sneezes. In one study, the virus was recovered in 1/13 of sneezes and 0/8 coughs generated by adults with natural rhinovirus (cold) infections.

The role of body cooling in causing the common cold is controversial. It is the most commonly offered folk explanation for the disease, and it has received some experimental evidence. One study showed that exposure to the cold causes cold symptoms in about 10% of those exposed, and that the subjects experiencing this effect report far more colds overall than those who do not. However, a variety of other studies do not show such an effect.

A history of smoking extends the duration of illness by about three days.

Getting fewer than seven hours of sleep per night has been associated with a risk three times higher of developing an infection when exposed to a rhinovirus, compared to those who sleep more than eight hours per night.

Common colds are seasonal, occurring more frequently during winter outside of tropical zones. Some argue that this is partly due to a change in behaviours such as increased time spent indoors, which puts infected people in close proximity to other people, rather than the exposure to cold temperatures.

Low humidity increases viral transmission rates. One theory is that dry air causes evaporation of water, thus allowing small viral droplets to disperse farther and stay in the air longer.

Counterintuitively, people with stronger immune systems are more likely to develop symptomatic colds. This is because the symptoms of a cold are directly due to the strong immune response to the virus, not the virus itself. People with less active immune systems—about a quarter of adults—get infected with the viruses, but the relatively weak immunological response produces no significant or identifiable symptoms. These people are asymptomatic carriers and can unknowingly spread the virus to other people. Because strong immune responses cause cold symptoms, "boosting" the immune system increases cold symptoms.

Pathophysiology

The common cold is a type of pharyngitis (inflammation of the throat). In the common cold, the inflammation is caused by a viral infection in the uppermost part of the throat (the nasopharynx), which runs from behind the nose down to the mouth. The common cold virus is transmitted mainly from contact with saliva or nasal secretions of an infected person, either directly, when a healthy person breathes in the virus-laden aerosol generated when an infected person coughs or sneezes, or by touching a contaminated surface and then touching the nose or eyes.

Symptoms are not necessary for viral shedding or transmission, as a percentage of asymptomatic subjects exhibit viruses in nasal swabs. It is generally not possible to identify the virus type through symptoms, although influenza can be distinguished by its sudden onset, fever, and cough.

The major entry point for the virus is normally the nose, but can also be the eyes (in this case drainage into the nasopharynx would occur through the nasolacrimal duct). From there, it is transported to the back of the nose and the adenoid area. The virus then attaches to a receptor, ICAM-1, which is located on the surface of cells of the lining of the nasopharynx. The receptor fits into a docking port on the surface of the virus. Large amounts of virus receptor are present on cells of the adenoid. After attachment to the receptor, virus is taken into the cell, where it starts an infection, and increases ICAM-1 production, which in turn helps the immune response against the virus. Rhinovirus colds do not generally cause damage to the nasal epithelium. Macrophages trigger the production of cytokines, which in combination with mediators cause the symptoms. Cytokines cause the systemic effects. The mediator bradykinin plays a major role in causing the local symptoms such as sore throat and nasal irritation.

The common cold is self-limiting, and the host's immune system effectively deals with the infection. Within a few days, the body's humoral immune response begins producing specific antibodies that can prevent the virus from infecting cells. Additionally, as part of the cell-mediated immune response, leukocytes destroy the virus through phagocytosis and destroy infected cells to prevent further viral replication. In healthy, immunocompetent individuals, the common cold resolves in seven days on average.

Prevention

The best prevention for the common cold is staying away from people who are infected, and places where infected individuals have been. Regular hand washing is recommended to reduce transmission of cold viruses and other pathogens via direct contact. Virus can be recovered from the hands of <“40% of adults with rhinovirus colds, and the quantity of virus recovered from the hands is also generally greater than that recovered in coughs and sneezes. Washing of the hands reduces virus count on the skin.

Hand washing with plain soap and water is recommended. The mechanical action of hand rubbing with plain soap, rinsing, and drying physically removes the virus particles from the hands.

Alcohol-based hand sanitizers provide very little protection against upper respiratory infections, especially among children.

Because the common cold is caused by a virus instead of a bacterium, antibacterial soaps are no better than regular soap for removing the virus from skin or other surfaces.

Aqueous iodine has been found to reliably eliminate the cold virus on human skin, however iodine is not acceptable for general use as a virucidal hand treatment because it discolours and dries the skin.

Randomised controlled trials have shown that hand washing using different combinations of cleaning agents resulted in a reduction in the incidence of rhinovirus infections. In two studies at day care facilities, increased handwashing of caregivers reduced the incidence of colds in children by up to 20%. However, scheduled handwashing at an elementary school has been shown to reduce the incidence of all communicable illnesses and gastrointestinal illnesses in particular, but was not shown to prevent respiratory ailments.

Efforts to develop a vaccine against the common cold have been unsuccessful. Common colds are produced by a large variety of rapidly mutating viruses; successful creation of a broadly effective vaccine is highly improbable. Exposure to cold temperatures and dry weather have been found to facilitate viral infection, explaining why colds and flu are more prevalent in winter outside of tropical areas. Cold weather may make the mucous lining of the respiratory tract more sluggish, taking longer to sweep any inhaled virus particles away. This allows more time for the virus to establish infection and means an individual is infectious (exhaling virus particles) for longer. In humidity above 80%, droplets containing viruses fall out of the air.

However, whilst it creates a better environment for the virus, cold weather itself does not directly cause colds and neither is there evidence supporting the idea that cold weather weakens the cells involved in the immune response.

Management

There are currently no medications or herbal remedies which have been conclusively demonstrated to shorten the duration of infection in all people with cold symptoms.

Treatment comprises symptomatic support usually via analgesics for fever, headache, sore muscles, and sore throat.

Symptomatic

Treatments that help alleviate symptoms include simple analgesics and antipyretics such as ibuprofen and acetaminophen/paracetamol. Evidence does not show that cough medicine is any more effective than simple analgesics and is not recommended for use in children due to a lack of evidence supporting its effectiveness and the potential for harm.

Symptoms of a runny nose can be reduced by a first generation antihistamine; however, it can cause drowsiness and other side effects. Other decongestants such as pseudoephedrine are effective in adults but there is insufficient evidence to support their use in children. Anticholinergics such as Ipratropium nasal spray can reduce the symptoms of runny nose with less side effects.

One study has found chest vapour rub to be effective at providing some symptomatic relief of nocturnal cough, congestion, and sleep difficulty.

Getting plenty of rest, drinking fluids to maintain hydration, and gargling with warm salt water, are reasonable conservative measures. Evidence for encouraging the active intake of fluids in acute respiratory infections is lacking as is the use of heated humidified air. Saline nasal drops may help alleviate nasal congestion.

Antibiotics and Antivirals

Antibiotics have no effect against viral infections and thus have no effect against the viruses that cause the common cold and due to their side effects cause overall harm. There are no approved antiviral drugs for the common cold even though some preliminary research has shown benefit.

Alternative Treatments

While many alternative treatments are used there is insufficient scientific evidence to support the use of most. Honey may be effective treatment in decreasing cough and improving sleep in children more than no treatment or dextromethorphan. However, honey should not be given to a child younger than one year old because of the risk of infant botulism. The benefits versus risk of nasal irrigation are currently unclear and therefore it is not recommended.

Zinc deficiency impairs immune function. It has been suggested that zinc may inhibit rhinovirus replication and reduce inflammation. Trials have found that zinc supplements can somewhat reduce the severity and duration of common cold symptoms when taken by otherwise healthy adults within 24 hours of onset of symptoms.

Vitamin C's effect on the common cold has been extensively researched. It has not been shown effective in prevention or treatment of the common cold, except in limited circumstances (specifically, individuals exercising vigorously in cold environments). Routine vitamin C supplementation does not reduce the incidence or severity of the common cold in the general population, though it may reduce the duration of illness.

Evidence about the usefulness of echinacea supplements, a popular herbal remedy, is contradictory. Well-conducted research studies tend to have negative results at a much higher rate than poorly conducted studies. Different types of echinacea supplements may vary in their effectiveness.

Generally, those studies that show supportive results indicate that echinacea might reduce the likelihood of developing cold symptoms upon inoculation with a virus by about half.

Prognosis

The common cold is generally mild and self-limiting.

Pneumonia is a possible complication.

Epidemiology

Upper respiratory tract infections are the most common infectious diseases among adults, who have two to four respiratory infections annually. Children may have six to ten colds a year (and up to 12 colds a year for school children).

In the United States, the incidence of colds is higher in the fall (autumn) and winter, with most infections occurring between September and April. The seasonality may be due to the start of the school year, or due to people spending more time indoors (thus in closer proximity with each other) increasing the chance of transmission of the virus.

History

The name "common cold" came into use in the 16th century, due to the similarity between its symptoms and those of exposure to cold weather. Norman Moore relates in his history of the Study of Medicine that James I continually suffered from nasal colds, which were then thought to be caused by polypi, sinus trouble, or autotoxaemia.

In the 18th century, Benjamin Franklin considered the causes and prevention of the common cold. After several years of research he concluded: "People often catch cold from one another when shut up together in small close rooms, coaches, etc. and when sitting near and conversing so as to breathe in each other's transpiration."

Although viruses had not yet been discovered, Franklin hypothesized that the common cold was passed between people through the air. He recommended exercise, bathing, and moderation in food and drink consumption to avoid the common cold. Franklin's theory on the transmission of the cold was confirmed some 150 years later.

Common Cold Unit

In the United Kingdom, the Common Cold Unit was set up by the Medical Research Council in 1946. The unit worked with volunteers who were infected with various viruses. The rhinovirus was discovered there. In the late 1950s, researchers were able to grow one of these cold viruses in a tissue culture, as it would not grow in fertilised chicken eggs, the method used for many other viruses. In the 1970s, the CCU demonstrated that treatment with interferon during the incubation phase of rhinovirus infection protects somewhat against the disease, but no practical treatment could be developed. The unit was closed in 1989, two years after it completed research of zinc gluconate lozenges in the prophylaxis and treatment of rhinovirus colds, the only successful treatment in the history of the unit.

Social and Cultural

Economics

In the United States, the common cold leads to 75 to 100 million physician visits annually at a conservative cost estimate of $7.7 billion per year. Americans spend $2.9 billion on over-the-counter drugs and another $400 million on prescription medicines for symptomatic relief.

More than one-third of patients who saw a doctor received an antibiotic prescription, which has implications for antibiotic resistance from overuse of such drugs.

An estimated 22 to 189 million school days are missed annually due to a cold. As a result, parents missed 126 million workdays to stay home to care for their children. When added to the 150 million workdays missed by employees suffering from a cold, the total economic impact of cold-related work loss exceeds $20 billion per year. This accounts for 40% of time lost from work.

Legal

Canada in 2009 restricted the use of over-the-counter cough and cold medication in children 6 years and under due to concerns regarding risks and unproven benefits.

Cold Weather

The traditional folk theory is that a cold can be "caught" by prolonged exposure to cold weather such as rain or winter conditions, which is where the disease got its name. Common colds are seasonal in temperate latitudes, with more occurring during winter. The experimental evidence for this effect is uneven: many experiments have failed to produce evidence that short-term exposure to cold weather or direct chilling increases susceptibility to infection, implying that the seasonal variation is instead due to a change in behaviours such as increased time spent indoors at close proximity to others. However, other experiments do find such an effect for both body chilling and cold air exposure, and a number of mechanisms by which lower temperatures could compromise the immune system have been suggested, while other experiments have shown that exposure to cold temperatures may instead stimulate the immune system.

Research

Dermanyssus Gallinae

Dermanyssus gallinae (also known as the red mite, poultry mite, red poultry mite and chicken mite) is an ectoparasite of poultry and other bird species.

Description

The mites are blood feeders and attack resting birds at night. They are generally white or greyish in colour, becoming darker or redder when engorged with blood. After feeding, they hide in cracks and crevices away from daylight, where they mate and lay eggs. The mite progresses through 5 life stages: egg, larva, protonymph, deutonymph and adult. Under favourable conditions this life cycle can be completed within seven days, so populations can grow rapidly - causing anaemia in badly affected flocks of poultry. Young birds are most susceptible. The mites can also affect the health of the birds indirectly, as they may serve as vectors for diseases such as Salmonellosis, avian spirochaetosis and *Erysipelothrix rhusiopathiae*.

Dermansyssus gallinae can also feed on some species of mammals, including humans, - causing dermatitis and skin lesions. However the mite needs an avian host to reproduce.

Clinical Signs and Diagnosis

The mites normally feed around the breast and legs of hens, causing pain, irritation, and a decrease in egg production. Pustules, scabs, hyperpigmentation and feather loss may develop.

If they are present in large numbers, *D. gallinae* can cause anaemia in hens which presents as pallor of the comb and wattle.

A presumptive diagnosis can be made in flocks of laying hens, usually based on a history of decreasing egg production, anaemia and mortalities in young or ill birds. Definitive diagnosis is only achieved following identification of eggs, faeces or the mites themselves

Treatment and Prevention

Ectoparasiticides can be used to treat affected poultry. These chemical controls, if used, should be used in rotation to avoid the build up of resistance. Red mites can survive for up to 10 months in an empty hen house. Creosote treatment of wood will kill mites.

The Poultry Red Mite, Dermanyssus Gallinae, A Potential Vector of Erysipelothrix Rhusiopathiae Causing Erysipelas in Hens

Egg Binding

In farming, aviculture and animal husbandry, the term egg binding refers to a medical condition in birds where the female is unable to pass an egg that has formed.

The egg may be stuck near the cloaca, or further inside. Egg binding is a reasonably common, and potentially serious, condition that can lead to infection or damage to internal tissue. The bound egg may be gently massaged out; failing this it may become necessary to break the egg *in situ* and remove it in parts. If broken, the oviduct should be cleaned of shell fragments and egg residue to avoid damage or infection. The term can also be seen in herpetoculture, as this condition can occur in female reptiles. It is inadvisable to attempt to break a reptile egg to remove it from an egg bound female. This procedure may be done by a veterinarian, who will insert a needle into the egg, and withdraw the contents with a syringe, allowing the egg to collapse and be removed. Nonsurgical interventions include administering oxytocin to improve contractions and allow the eggs to pass normally. In many cases, egg bound reptiles must undergo surgery to have stuck eggs removed.

Egg binding in reptiles is quickly fatal if left untreated, therefore gravid females who become very lethargic and cease feeding, need immediate medical treatment in order to treat the potentially life-threatening condition. A recent episode of the Animal Planet reality show E-Vet Interns featured the treatment of an egg bound turtle named Napoleon. Exotics specialist Dr. Kevin Fitzgerald of Alameda East Veterinary Hospital is shown treating her with oxytocin and then eventually having to resort to surgery with footage of the large number of eggs that were removed. Dr. Fitzgerald was shown explaining to the new interns how dangerous this condition can be for a pet turtle and the need for early medical intervention.

Egg binding can occur if an egg is malformed and/or too large, the animal is weakened by illness, improper husbandry, or stress, or if hormonal balances are wrong (producing weak contractions). Factors that can contribute to the risk of egg binding include calcium deficiency, breeding animals that are too young or too small, not providing suitable laying areas (leading to deliberate retention of eggs), and overfeeding of species in which clutch size is dependent on food intake (such as Veiled Chameleons).

Egg Binding

You first generally notice this ailment rather too late, when it is obvious the hen is in some discomfort. The hen often stands or moves in an odd way, usually with her tail held very low and her rear end tucked between her legs. Sometimes they just sit around looking ruffled, but often it is obvious the bird is straining to pass an egg. If you spot it early your chances of helping the hen are greatest.

Egg binding can happen in young hens just starting to lay or in older birds that have become fat. Lack of exercise can cause fat to build up in the birds body, around the reproductive organs and so cause the egg to get stuck. Lack of calcium in the diet can be a major cause of it. Sometimes the egg is just too large for the bird to pass, sometimes the shell is rough and not easily expelled.

In my experience the egg bound hen, (Unless it's a pullet first starting to lay, these seem to respond better) has not got a very good outlook. If you approach the problem with this in mind then anything you can do to save your bird is a bonus. Unless the egg gets passed without too much fuss, it frequently seems to cause the bird massive shock, the bird often will die. Egg binding can also cause a prolapse, which will forever cause problems with that hen afterwards. However going on the principle that if its going to die without help. You have to try something!

So here are some things you can do to assist the hen, and are always worth trying, especially on a young bird who is just having trouble with her first egg. These birds are usually the most rewarding to treat, they seem to respond better and often it does not reoccur with them, unlike older birds who have developed internal problems.

The first thing to do is to put the bird somewhere warm for a while, often, this treatment in itself can help enormously. As with many bird related ailments, heat can be a wonderful healer. If the hen is in shock from it is vital. A comfortable heat will often give the bird enough of a boost to be able to pass the egg herself. Hopefully this will be the case with your bird. If after a while she is still straining and no egg has arrived, I would suggest gently introducing some slightly warmed oil (body temp) into the vent, cooking oil is fine. If the egg is visible hold the hen vent downwards over a bowl of gently steaming water.. (Don't over-heat the poor thing she's going through hell as it is!)

Many people will tell you that if you break the egg the hen will die. Yes this is very often the case, sometimes the act of breaking the egg may cause the bird to just have a heart attack and drop dead. I've seen it happen, but I have had success on two occasions using this method. I gently and carefully made a small hole in the visible end of the egg, and emptied the contents. The contractions of the hen quickly crushed the now empty shell and I pulled it free. DO NOT leave any egg behind!

If your hen finally lays her egg she will show immediate and understandable relief! I always give the hen some electrolytes to drink and a light feed and usually they are happy to go back to the flock after a brief rest. In the case of an older bird, I usually put her on a light diet for a few days to try to bring her out of lay. Especially if I think it is due to the hen being overweight.

Now for some modern stuff that I have heard about but haven't tried, but I'm absolutely going to ask my vet about! I have recently heard that excellent results can be obtained very quickly with injectable calcium gluconate which is given as an inter muscular injection into the pectoral muscle. I want to find out more about this as it could be very useful in future.

Preventing Egg Binding

I have found that by using a calcium/mineral supplement added to the birds layers rations a couple of times a week that this problem has decreased. I have a lot of birds and generally used to expect to have a couple of cases of egg binding during the season. The last couple of years its been very minor, if at all, so I think that the added calcium/minerals may have just given that extra help needed to help prevent this ailment. As calcium is required for the muscle contractions which push the egg out of the body as well as for the formation of the shell I think extra calcium and minerals are a very good idea during the breeding season. I for one will not rely on breeders pellets and oyster shell alone.

Fatty Liver Hemorrhagic Syndrome

Fatty liver hemorrhagic syndrome (also referred to as fatty liver syndrome), a disease in chickens and other birds, affects only hens (females). Birds with this disease have large amounts of fat deposited in their liver and abdomen.

This often results in an enlarged liver that is easily damaged and prone to bleeding. In some cases the disease is fatal, usually as a result of blood loss from an internal hemorrhage in the liver. The hemorrhage often occurs when a hen is straining to lay her egg. Fatty liver hemorrhagic syndrome is "the major cause of mortality in laying hens."

Causes

Excessive dietary energy intake is believed to be the cause of fatty liver hemorrhagic syndrome. Heredity may also play a role, but it is not the entire cause for the disease. Birds housed in cages will more likely be affected because they are unable to exercise to burn off the extra dietary energy.

Walking hens are less likely to develop this problem. The disease is observed most often in birds that appear to be healthy and in a state of high egg production. As a result, death can occur quite unexpectedly.

Symptoms

Affected birds are usually overweight and may also have pale combs. Generally, however, the disease has few or no symptoms prior to the bird's death.

Treatment

The use of L-Tryptophan in the diet can decrease the syndrome.

Fowlpox

Fowlpox is a worldwide disease of poultry caused by viruses of the family *Poxviridae* and the genus *Avipoxvirus*. The viruses causing fowlpox are distinct from one another but antigenically similar, possible hosts including chickens, turkeys, quail, canaries, pigeons, and many other species of birds.

There are two forms of the disease. The first is spread by biting insects (especially mosquitoes) and wound contamination and causes lesions on the comb, wattles, and beak. Birds affected by this form usually recover within a few weeks. The second form is spread by inhalation of the virus and causes a diphtheritic membrane to form in the mouth, pharynx, larynx, and sometimes the trachea. The prognosis for this form is poor.

Vaccines are available for fowlpox (ATCvet code: QI01AD12). Chicken are usually vaccinated with *pigeonpox virus*. Turkeys are also routinely vaccinated.

Turkeypox

Turkeypox is a virus of the family *Poxviridae* and the genus *Avipoxvirus*. It is one of the most common diseases in the Wild Turkey (Meleagris gallopavo) population.

Causes

Turkeypox is caused by the fowlpox virus, which infects birds.

Turkeypox, like all avipoxviruses, is transmitted either through skin contact or by arthropods (typically mosquitos) acting as mechanical vectors.

Gallid Herpesvirus 1

Gallid herpesvirus 1 (GaHV-1) (also known as Avian herpesvirus 1) is a virus of the family *Herpesviridae* that causes avian infectious laryngotracheitis. It was originally recognised as a disease of chickens in the United States in 1926. The disease also occurs in pheasants. The disease is usually referred to as Infectious laryngotracheitis or simply LT in the poultry industry.

It is widely viewed as one of the most contagious viruses that affect the poultry industry. A confirmed case will usually result in the establishment of a quarantine zone around the farm. Inside this quarantine zone, poultry workers will avoid poultry farms to prevent the spread of the virus.

GaHV-1 is shed in respiratory secretions and transmitted by droplet inhalation or via fomites. A previously unexposed flock will develop cases for two to eight weeks following introduction. The incubation period is two to eight days.

Clinical Signs and Diagnosis

Symptoms include coughing, sneezing, head shaking, lethargy, discharge from the eyes and nostrils (sometimes bloody), and difficulty breathing. The name comes from the severe inflammation of the larynx and trachea. A diphtheritic membrane may form in the trachea, causing obstruction.

There may be problems in egg laying and the production of abnormal or thin-shelled eggs.

Mortality is typically less than 15 percent.

Histopathology, PCR, ELISA, immunofluorescent staining and viral isolation are all possible methods of diagnosis.

Treatment and Control

A vaccine is available (ATCvet code: QI01AD08), but it does not prevent latent infections. It can be used during an outbreak to decrease morbidity and deaths.

Biosecurity measure including quarantine, isolation and disinfection are very important in controlling the spread of an outbreak.

7

Avian Infectious Laryngotracheitis

Introduction

Gallid Herpes virus causes respiratory disease in chickens and pheasants.

Disease varies from mild to peracute, with mortality in peracute outbreaks exceeding 50%. As with all herpesviruses, GHV-1 can remain latent in carriers after infection and then be shed intermittently, recrudescing with stress.

Signalment

The chicken is the primary host and reservoir host. A form of LT has been described in pheasants.

Distribution

Worldwide. Transmission is via direct contact and contaminated people and equipment. Vermin and wild birds and dogs may aid mechanical transmission.

Clinical Signs

Respiratory signs:

- Nasal discharge which is often bloody
- Coughing which may also include blood
- Sneezing, dyspnoea, gasping, upper respiratory tract pain
- Abnormal lung sounds
- Decreased egg production, thin egg shells, lack of growth
- Neurological and ophthalmologic signs may develop.

Death may occur rapidly and with high mortality in peracute and acute disease. In recent times, LT usually presents in a mild form and most birds recover.

Diagnosis

On post-mortem, haemorrhagic tracheitis and bloodstained mucus are evident. Pneumonia and sacculitis may also be seen. Caseous diptheritic membranes may be present on the mucosae of the upper respiratory tract. Histopathology reveals loss of cilia, mucosal gland atrophy, intranuclear inclusion bodies and epithelial cell sloughing. Characteristic syncytia develop. A fibrinonecrotic membrane may be present in more chronic disease cases.

Antigen ELISA is both straightforward, quick and sensitive. The PCR can be used to detect LTV. Immunofluorescent or Immunoperoxidase staining can also be performed and is more rapid but less sensitive. Virus isolation on a variety of tissues including tracheal swabs or tissue samples may be useful.

Agar Gel Immunodiffusion can detect virus in tracheal samples.

Electron microscopy can be used to demonstrate viral particles in tracheal scrapings or exudates but is insensitive. Measuring viral antibody measures infection indirectly as serum antibodies peak around 2 weeks after infection and wane slowly afterwards.

Treatment

Where early diagnosis is made, vaccination can be administered in the face of infection to help reduce further morbidity and mortality.

Control

ILT can be effectively controlled by vaccination. Vaccinated and unvaccinated birds should not be mixed due to the possibility of reversion to virulence. Most are modified live isolates and are administered by eye drop.

Adequate Biosecurity, Quarantine and Disinfection is also Essential. Wild birds and vermin should be prevented from accessing poultry and their food/water sources.

Gapeworm

A gapeworm (*Syngamus trachea*) is a parasitic nematode worm infecting the tracheas of certain birds. The resulting disease, known as gape or the gapes, occurs when the worms clog and obstruct the airway. The worms are also known as red worms or forked worms due to their red colour and the permanent procreative conjunction of males and females. Gapeworm is common in young, domesticated chickens and turkeys.

When the female gapeworm lays her eggs in the trachea of an infected bird, the eggs are coughed up, swallowed, then defecated. When birds consume the eggs found in the feces or an intermediate host such as earthworms, snails (*Planorbarius corneus*, *Bithynia tentaculata*, ...), or slugs, they become infected with the parasite.

Ivermectin is a drug often used to control gapeworm infection in birds.

Morphology

Males and females are joined together in a state of permanent copulation forming a Y shape "forked worm". They are also known as the "red worm" because of their colour. Females are much larger (up to 20 mm long) than males (up to 6 mm long). The life history of the gapeworm is peculiar in that transmission from bird to bird may be successfully accomplished either directly (by the feeding of embryonated eggs or infective larvae) or indirectly (by ingestion of earthworms containing free or encysted gapeworm larvae they had obtained by feeding on contaminated soil).

Life Cycle and Pathogenesis

The life cycle is complicated in both its preparasitic and parasitic phases. In the preparasitic phase, L3s develop inside the eggs at which time they may hatch. Earthworms play an important role in the life cycle, serving as transport (paratenic) hosts. Larvae have been shown to remain viable for more than three years encapsulated in earthworm muscles. Other invertebrates may also serve as paratenic hosts and these include terrestrial snails and slugs as well as the larvae of Musca domestica (the common house fly) and Lucilia sericata (the green bottle fly responsible for cutaneous myiasis).

The parasitic phase involves substantial migration in the definitive host to reach the predilection site. Young birds are most severely affected with migration of larvae and adults through the lungs causing a severe pneumonia. Lymphoid nodules form at the point of attachment of the worms in the bronchi and trachea. Adult worms also appear to feed on blood. Worms in the bronchi and trachea provoke a hemorrhagic tracheitis and bronchitis with formation of large quantities of mucus, plugging the air passages and, in severe cases, causing asphyxiation. Pheasants appear to be particularly susceptible to infections resulting in mortality rates as high as 25% during outbreaks. The rapidly growing worms soon obstruct the lumen of the trachea and cause suffocation. Turkey poults, baby chicks and pheasant chicks are most susceptible to infection. Turkey poults usually develop gapeworm signs earlier and begin to die sooner after infection than young chickens. Lesions are usually found in the trachea of turkeys and pheasants but seldom if ever in the tracheas of young chickens and guinea fowl. The male worm, in the form of lesions, remains permanently attached to the tracheal wall throughout the duration of its life. The female worms apparently detach and reattach from time to time in order to obtain a more abundant supply of food.

Epidemiology

Earthworm transport hosts are important factors in the transmission of Syngamus trachea where poultry and game birds are reared on soil. The longevity of L3s in earthworms (up to 3 years) is particularly important in perpetuating the infection from year to year. Wild birds may serve as reservoirs of infection and have been implicated as the sources of infections in outbreaks on game-bird farms as well as poultry farms. Wild reservoir hosts may include pheasants, ruffed grouse, partridges, turkeys, magpies, meadowlarks, robins, grackles, jays, jackdaws, rooks, starlings and crows.

There is also evidence to suggest that strains of Syngamus trachea from wild bird reservoir hosts may be more infective for domestic birds if they first pass through an earthworm transport host rather than by direct infections via ingestion of L3s or eggs containing L3s. Clinical signs Blockage of the bronchi and trachea with worms and mucus will cause infected birds to gasp for air.

They stretch out their necks, open their mouths and gasp for air producing a hissing noise as they do so. This "gaping" posture has given rise to the common term "gapeworm" to describe Syngamus trachea. These clinical signs first appear approximately 1-2 weeks after infection. Birds infected with gapeworms show signs of weakness and emaciation and usually spend much of their time with eyes closed and head drawn back against the body. An infected bird may give its head a convulsive shake in an attempt to remove the obstruction from the trachea so that normal breathing may be resumed. Severely affected birds, particularly young ones, will deteriorate rapidly; they stop drinking and become anorexic. At this stage, death is the usual outcome. Adult birds are usually less severely affected and may only show an occasional cough or even no obvious clinical signs.

Diagnosis

A diagnosis is usually made on the basis of the classical clinical signs of "gaping". Subclinical infections with few worms may be confirmed at necropsy by finding copulating worms in the trachea and also by finding the characteristic eggs in the feces of infected birds. Examination of the trachea of infection birds shows that the mucous membrane is extensively irritated and inflamed. Coughing is apparently the result of this irritation to the mucous lining.

Control and Treatment of Syngamus Trachea

Prevention

In the artificial rearing of pheasants, gapes are a serious menace. Confinement rearing of young birds has reduced the problem in chickens compared to a few years ago. However, this parasite continues to present an occasional problem with turkeys raised on range. Confinement rearing of broilers/pullets and caging of laying hens, have significantly influenced the quantity and variety of nematode infections in poultry. For most nematodes, control measures consist of sanitation and breaking the life cycle rather than chemotherapy. Confinement rearing on litter largely prevents infections with nematodes using intermediate hosts such as earthworms or grasshoppers, which are not normally found in poultry houses. Conversely, nematodes with direct life cycles or those that utilise intermediate hosts such as beetles, which are common in poultry houses, may prosper.

Treatment of the soil or litter to kill intermediate hosts may be beneficial. Insecticides suitable for litter treatment include carbaryl, tetrachlorvinphos (stirofos). However, treatment is usually done only between grow-outs. Extreme care should be taken to ensure that feed and water are not contaminated. Treatment of range soil to kill ova is only partially successful. Changing litter can reduce infections, but treating floors with oil is not very effective. Raising different species or different ages of birds together or in close proximity is a dangerous procedure as regards parasitism. Adult turkeys, which are carriers of gapeworms, can transmit the disease to young chicks or pheasants, although older chickens are almost resistant to infection.

Treatment

Thiabendazole (Tresaderm) is currently approved for use only in pheasants and is effective when administered in the feed. Continuous medication of pen-reared birds has been recommended, but is not economical. Several other compounds have been shown effective against S. trachea under experimental conditions.

Methyl 5-benzoyl-2-benzimidazole was 100% efficacious when fed prophylactically to turkey poults. 5-isopropoxycarbonylamino-2-(4-thizolyl)-benzimidazole was found to be more efficacious than thiabendazole or disophenol.

The level of control with three treatments of cambendazole on days 3-4, 6-7, and 16-17 post-infection was 94.9% in chickens and 99.1% in turkeys. Levamisole (Ergamisol), fed at a level of 0.04% for 2 days or 2 g/gal drinking water for 1 day each month, has proven effective in game birds. Fenbendazole (Panacur) at 20 mg/kg for 3-4 days is also effective.

Infectious Bursal Disease

Etiology and Transmission

Infectious bursal disease is caused by a birnavirus (IBDV) that is most readily isolated from the bursa of Fabricius but may be isolated from other organs. It is shed in the feces and transferred from house to house by fomites. It is very stable and difficult to eradicate from premises.

IBDV may be isolated in 8 to 11 day old, antibody-free chicken embryos with inocula from birds in the early stages of disease. The chorioallantoic membrane is more sensitive to inoculation than is the allantoic sac. IBDV also may be isolated in cell cultures derived from the cloacal bursa and established cell lines, and some strains may be isolated in chicken-embryo fibroblasts.

Cell-culture-adapted strains of IBDV produce a cytopathic effect and may be used for quantitative serologic tests. Two serotypes of IBDV have been identified; within them, antigenic variation between strains is considerable. Serotype 2 infects chickens and turkeys but does not cause clinical disease or immunosuppression.

"Variant" strains of IBDV, which have major antigenic differences from the "standard" strains, cause immunosuppression but not clinical disease in older chickens.

Clinical Findings

Infectious bursal disease is highly contagious; results of infection depend on age and breed of chicken and virulence of the virus. Infections may be subclinical or clinical. Infections before 3 wk of age are usually subclinical. Chickens are most susceptible to clinical disease at 3-6 wk, but severe infections have occurred in Leghorn chickens up to 18 wk old.

Early subclinical infections are the most important form of the disease because of economic losses. They cause severe, long-lasting immunosuppression due to destruction of immature lymphocytes in the bursa of Fabricius, thymus, and spleen. The humoral (B cell) immune response is most severely affected; the cell-mediated (T cell) immune response is affected to a lesser extent.

Chickens immunosuppressed by early IBDV infections do not respond well to vaccination and are predisposed to infections with normally nonpathogenic viruses and bacteria. Common diseases are usually exacerbated by IBDV infections. Subclinical infections by the "variant" strains occur in immature birds, and severe longterm immunosuppression and bursal atrophy result from early infections.

In clinical infections, onset of the disease is sudden after an incubation of 3-4 days. Chickens exhibit severe prostration, incoordination, watery diarrhea, soiled vent feathers, vent picking, and inflammation of the cloaca.

Losses range to >20%. Recovery occurs in <1 wk, and broiler weight gain is delayed by 3-5 days. The presence of maternal antibody will modify the clinical course of the disease. Virulence of field strains of the virus varies considerably. Very virulent (vv) strains of the virus that cause high mortality and morbidity were detected first in Europe. These spread throughout the Old World in the last decade and in 1999 were in South America. The vv strains have not been detected in the USA.

Lesions

At necropsy, the cloacal bursa is swollen, edematous, yellowish, and occasionally hemorrhagic, especially in birds that have died of the disease. Congestion and hemorrhage of the pectoral, thigh, and leg muscles is common. Chickens recovered from IBDV infections have small, atrophied, cloacal bursas due to the destruction and lack of regeneration of the bursal follicles.

Control

There is no treatment. Depopulation and rigorous disinfection of contaminated farms have achieved limited success. Live vaccines of chick-embryo or cell-culture origin and of varying virulence can be administered by eye drop, drinking water, or SC routes at 1-21 days of age. The immune response can be altered by maternal antibody, and the more virulent vaccine strains can override higher levels of antibody.

High levels of maternal antibody during early brooding of chicks in broiler flocks (and in some commercial layer operations) can minimise early infection, subsequent immunosuppression, or both. Breeder flocks should be vaccinated one or more times during the growing period, first with a live vaccine and again just before egg production with an oil-adjuvanted, inactivated vaccine. Inactivated vaccines of chick-embryo, bursa, or cell-culture origin are available. The latter vaccines induce higher, more uniform, and more persistent levels of antibody than do live vaccines. The immune status of breeder flocks should be monitored periodically with a quantitative serologic test such as virus neutralisation or ELISA. If antibody levels fall, hens should be revaccinated to maintain adequate immunity in the progeny.

Bloodborne Organisms

Avian blood may contain various disease agents including viruses, bacteria, rickettsiae, protozoa, microfilariae, and rarely fungi. Except for viruses, these organisms often can be identified by microscopic examination of wet mounts, buffy coat, or blood smears; appropriate culturing techniques; or subinoculation of blood into susceptible birds. Microscopically, some are within blood cells (Plasmodium, Haemoproteus, Leucocytozoon, Atoxoplasma, Hepatocystis, Babesia, Aegyptianella), while others are free in the plasma (Trypanosoma, microfilariae, bacteria, spirochetes). None lives exclusively in the blood; most are found in tissues but are present in blood during part of their life cycle. Some, such as microfilariae and Plasmodium, may have a periodicity when numbers or stages of parasites are present at different times. In such cases, examining multiple smears at intervals will increase the likelihood of obtaining a diagnosis. Seasonal variations in infection rates relate to the activity of arthropod vectors. When possible, tissue cytology is also a useful adjunct to examination of blood. Several bloodborne organisms are either uncommon or not associated with clinical disease. However, weakened or injured raptors infected with hemoprotozoa had higher mortality and delayed recovery compared with uninfected birds. Routine examination for bloodborne organisms should be included in the clinical and diagnostic procedures for any ill bird.

Thin blood smears should be made with blood directly from the bird if possible. Anticoagulants, storage, and cooling of the blood can distort protozoan morphology and introduce artifacts. A small drop of blood can be collected using a syringe and needle, or by selecting a small vessel in the wing web and, after cleaning the site thoroughly with alcohol and letting it dry, puncturing the vessel with a lancet so that a small drop of blood wells up from the wound. The drop should be picked up without touching the skin and spread on a clean glass slide at a 30° angle to make a thin smear. A good quality Romanowsky-type stain that gives good polychromatic coloration (eg, Giemsa stain) should be used. At least 200 oil-immersion fields (~20,000 RBC) for single smears or 100 for multiple smears from the same bird should be examined. Leucocytozoon and microfilariae are found around the periphery of smears and can be easily seen on low-power magnification.

Bloodborne organisms in plasma or WBC are concentrated in the buffy coat. A diamond-tipped pencil can be used to cut the microhematocrit tube just below the buffy coat above the packed RBC. The buffy coat should be expressed from the cut end with a small amount of plasma to make a suspension, and a thin smear prepared. Stained buffy coat smears are recommended for detecting bacteremia, spirochetes, and chronic Leucocytozoon, Trypanosoma, or Atoxoplasma infections. An excellent technique for identifying low numbers of motile organisms such as spirochetes and microfilariae is direct examination of the buffy coat by darkfield or phase contrast microscopy. The buffy coat and all of the plasma should be expressed onto a glass slide, and covered with a coverglass, which is depressed slightly to spread the buffy coat. The buffy coat/plasma interface should be examined with darkfield or reduced light microscopy to detect motile organisms.

Plasmodium infections can be determined by subinoculation. Ideally, birds of the same or a known susceptible avian species should be used, but this is often not practical. In general, canaries are used for detecting passerine infections, and turkeys are susceptible to most plasmodia that infect gallinaceous birds. Parasites remain viable in blood stored at 32°F (in ice) for at least 7 days. Inoculation IV is preferred and will result in earlier parasitemia, but any parenteral route can be used. Recipients should be examined twice weekly for a minimum of 4 wk if exposed IV; longer times are needed if other routes of inoculation are used. Spirochete, Aegyptianella, and bacterial infections can also be detected by subinoculation of infectious blood; bacteria can usually be identified by blood culture.

To make a diagnosis of infection with an intracellular blood protozoan on a thin blood film, it first should be determined that the "parasites" in question are neither normal nor artifact. The following should then be determined: the host cell and whether it is normal or deformed beyond identification, whether pigment granules are present or absent, and whether merogony is occurring. Identification of an organism beyond genus (or subgenus in the case of Plasmodium spp) is difficult and usually unnecessary for clinical purposes.

Aegyptianellosis

Aegyptianellosis is an acute, tickborne, febrile disease caused by Aegyptianella spp, a rickettsia in the family Anaplasmataceae. Infection of avian species, including chickens, turkeys, guineafowl, quail, pigeons, crows, waterfowl, ratites, passerines, and psittacines has been described. Ticks, especially Argas spp, transmit the organism; infection can also be reproduced by blood inoculation. Organisms appear as single or multiple, round, "signet-ring" (0.5-4 µm) or irregular oval bodies in RBC often lateral to the nucleus. Infections are most common in tropical and subtropical areas of Africa, Asia, and Europe; infection of wild turkeys in Texas has also been reported.

In endemic areas, infection is mild or asymptomatic. Ruffled feathers, anorexia, droopiness, diarrhea, fever, jaundice, and high mortality in younger birds occur in introduced or otherwise susceptible birds. Anaemia, which can lead to right heart failure and ascites, enlargement of the liver and spleen, enlarged discoloured kidneys, and pinpoint serosal hemorrhages are seen. Infestation with larval argasid ticks and Borrelia infection (spirochetosis, *Avian Spirochetosis: Introduction*) may accompany the disease.

Tetracyclines, especially doxycycline, are effective in controlling the disease and eliminating the organism from chronically infected birds. Tick control is an important adjunct to treatment.

Atoxoplasmosis (Isosporosis, Lankesterellosis)

Atoxoplasma are pale-staining, nonpigmented, oval, intracytoplasmic bodies within mononuclear cells currently thought to be lymphocytes. Usually, cells contain a single parasite, but multiple organisms can be seen in severe, acute infections. Presence of the protozoan causes the nucleus to curve around it giving the appearance that the organism is located within an indentation of the nucleus. At least 2 different genera of coccidian protozoa (Isospora, Lankesterella) have merozoites in the lymphocytes that are indistinguishable from each other and have been called Atoxoplasma. Passerine birds, especially canaries, finches, sparrows, and species of the Sturnidae family (starlings, mynahs) are affected by both protozoa. Poultry are not known to be affected.

Isospora has a direct life cycle that includes an extra-intestinal, systemic phase in some species. Merogony occurs in the intestine and oocysts are passed in the feces. The systemic part of the life cycle was initially not recognised, leading to the description of the intestinal stages and oocysts as Isospora serini. Transmission is faecal-oral via ingestion of oocysts in droppings from infected birds. Infected canaries can shed oocysts for at least 2 yr. Most species of avian Isospora do not have extra-intestinal stages. In canaries, I serini has a systemic phase while I canaria infects only the intestinal tract.

Lankesterella has stages in the blood indistinguishable from those caused by systemic Isospora spp, but merogony occurs in tissues other than the intestine, especially the lungs. The life cycle is indirect. Bloodsucking arthropods, particularly mites, become infected during feeding and transmit the protozoan when it refeeds on a susceptible bird.

Signs include listlessness, diarrhea, and anorexia. Mortality can be high (up to 80%) in young birds. In acutely affected birds, there is marked hepatomegaly and splenomegaly, often with multifocal necrosis. The enlarged liver and gallbladder can be seen through the abdominal wall, especially if it is moistened with alcohol, which provides the basis for the common name black spot disease. High numbers of parasites infecting lymphocytes are present in blood and organ impression smears. Nearly spherical oocysts averaging 19 × 21 μm are present in droppings of canaries infected with Isospora. Oocysts of I serini need to be distinguished from those of I canaria, which are slightly larger and more oval.

Diagnosis is difficult in chronically infected older birds. Very few parasites are present in blood and tissues, and oocysts are shed intermittently, although sometimes in high numbers. Buffy coat and organ smears are preferred. Negative findings should not be interpreted to mean that infection is not present. Hepatic and splenic enlargement persists because of infiltrations with high numbers of large lymphoid cells that serve as host cells for the parasite. Histopathologically, organisms are difficult to find and identify. Lesions may be mistaken for lymphosarcoma.

There is no known effective treatment. Anticoccidial drugs do not affect parasites in the tissues. Use of antimalarial drugs has been suggested. Good management procedures, including prevention of exposure to mites, isolation of age groups, and scrupulous cleanliness (particularly daily cleaning before oocysts sporulate) help control the disease. Disinfectants have little effect on oocysts.

Filariasis

Microfilariae are commonly found in the blood of wild birds but are rare to absent in poultry except in southeast Asia where infections in chickens and waterfowl occur. A high percentage of imported cockatiels have microfilariae. At least 16 genera of filarids are found in avian species. All have an indirect life cycle with bloodsucking insects (eg, lice, mosquitos, midges) serving as intermediate hosts.

Adults are relatively short lived and mature in body cavities, including the eye and ventricles of the brain, respiratory system, cardiovascular system, or connective tissues; some produce characteristic subcutaneous nodules. In contrast, microfilariae are long lived and may be numerous in the skin as well as in the circulation. Increased numbers of microfilariae have been seen in stressed individuals, but they rarely cause clinical disease or mortality. A possible exception is infection of emus with Chandlerella, a common filarid of the brain of free-living grackles. Parasites apparently do not produce microfilariae iemus. Affected emus show signs of CNS disease. Treatment with ivermectin, levamisole, or injection of nodules with 0.5% potassium permanganate solution, and surgical removal of adults have been used.

Haemoproteus Infection

Infections with Haemoproteus spp are common in nondomestic birds. Pigeons and doves are frequently infected. Species are found in free-living ducks, quail, and turkeys but are rare to absent in commercial flocks probably because of very specific feeding habits of Culicoides spp and hippoboscid flies, the invertebrate vectors. In free-living bird populations, females are more frequently infected than males. Until recently, Haemoproteus was considered relatively innocuous and of little clinical significance. Fatal infections can occur because of extensive widespread necrosis accompanying development of large exoerythrocytic megalomeronts in muscle, heart, liver, and lung. Mortality as high as 78% has occurred in bobwhite quail infected with Isospora lophortyx. Similar meronts and lesions have been reported previously in cases of "aberrant leucocytozoonosis" and arthrocystosis.

The diagnostic presence of large, pigmented gametocytes in mature RBC that often partially or completely encircle the nucleus without merogony can follow or occur simultaneously with systemic involvement. Sudden death without clinical signs or a prolonged course of weakness, lameness, dyspnea, lethargy, poor growth, and anaemia may be seen. Little is known about effective treatment, although antimalarial drugs may be tried. Chloroquine (5 mg/kg) and buparvaquone (2.5 mg/kg) have been reported to be effective in treating pigeons; diminazene and quinapyramine were either ineffective or toxic. Measures to control invertebrate hosts should help prevent heavy infections.

Leucocytozoonosis

Infections with Leucocytozoon spp range from subclinical to fatal. Mortality may approach 100% but varies greatly with species and strain of parasite, host species, degree of exposure, and other factors. Acute outbreaks of leucocytozoonosis have been reported in chickens (Asia, Africa), turkeys (North America), waterfowl (North America, Europe), and a number of free-living and captive avian species throughout the world. Species in domestic birds include L simondi in waterfowl; L smithi in turkeys; and L caulleryi, L sabrazesi, L andrewsi, and L schoutedeni in chickens.

L caulleryi can be highly pathogenic, causing a lethal hemorrhagic disease of chickens in southeast Asia. Numerous Leucocytozoon spp infect nondomestic birds (eg, blood smears from raptors often contain gametocytes). Clinical disease and mortality result from anaemia caused by antierythrocytic factors produced by the parasite, high numbers of the large gametocytes blocking pulmonary capillaries, or parasites invading the endothelium of vessels in vital tissues (brain, heart, etc.) where they form megalomeronts that occlude vessels and result in multifocal necrosis.

Wild birds are reservoirs in some areas and are responsible for initiating infection in young birds each year. Parasitemia often increases dramatically in late April and early May (called spring rise), just before arthropod vectors, black flies (Simulium spp), or biting midges (Culicoides spp) increase. Ducks that have recovered from infection with L simondi relapse when light cycles are manipulated to increase egg production. Increased levels of prolactin have been suggested as a possible cause. Acute disease is seen more often in the young when they have high parasitemia and when black flies or biting midges are most abundant. Subacute or chronic disease is seen in the young outside fly season and in older birds at any season; parasitemia is usually low. Recovered birds remain carriers and serve as a reservoir for young, susceptible birds.

Clinical Findings, Lesions, and Diagnosis

Acutely affected birds are listless and have anaemia, leukocytosis, tachypnea, anorexia, diarrhea with green droppings, and often CNS signs. Egg production is impaired in laying chickens infected with L caulleryi. Signs are evident ~1 wk after infection and coincide with the onset of parasitemia. Visibly affected birds die after 7-10 days or may recover with sequelae of poor growth and egg production. Hemorrhages, splenomegaly, and hepatomegaly are seen. Grossly visible white dots in affected organs are megalomeronts.

In thin blood smears, gametocytes may be seen along the edges and tail of the smear. Leucocytozoon is identified by large gametocytes that lack pigment and distort the host cell (RBC or WBC), making it no longer identifiable. Shape of gametocytes varies with the species—some are elongated with long tapering extremities, while others are round. Serology may detect prior infection.

Treatment and Control

Treatment usually is not effective. Preventive medication with combined pyrimethamine (1 ppm) and sulfadimethoxine (10 ppm) combined in the feed controls L caulleryi; clopidol (0.0125-0.025%) controls L smithi. Measures to control invertebrate vectors are helpful. Humoral immunity resulting from vaccination will protect against L caulleryi infection.

Plasmodium Infection

Plasmodium spp, which often are not host-specific, infect a wide variety of domestic and wild birds in most areas of the world and can cause high losses. P gallinaceum infects chickens in Asia and Africa and causes low mortality in indigenous chickens but rates may be as high as 80-90% in commercial birds. P juxtanucleare infects chickens in Asia, Africa, and South America; most infections are mild or asymptomatic. P durae infects turkeys and gallinaceous birds other than chickens in Africa; mortality in turkeys can approach 100%. Clinical malaria has not been reported from poultry in North America, but indigenous wild turkeys can become infected with at least 4 different Plasmodium species. Asymptomatic infections in endemic or introduced birds can be spread via mosquitos and cause fatal disease in introduced (eg, zoo penguins) or resident (eg, Hawaiian avifauna) birds, respectively. Invertebrate hosts are ornithophilic mosquitos, usually Culex, Culiseta, or Aedes spp.

Clinical Findings, Lesions, and Diagnosis

Infection with Plasmodium spp may be nonclinical or cause illness characterised by weakness, lassitude, dyspnea, anaemia, abdominal distention, increased right heart weight, ocular hemorrhage, and death. Death results from severe anaemia or blockage of capillaries in the brain or other vital organs by exoerythrocytic meronts in endothelial cells. Liver and spleen are markedly enlarged and often discoloured (dark brown to black). Pigmented parasites including meronts are found in both immature and mture RBC. Infrequently, parasites are found in thrombocytes and WBC. In birds that die acutely, organisms may be sparse or absent in blood, but numerous meronts can be found in capillaries by examining squash or impression smears of brain, lung, liver, and spleen. Serologic and molecular diagnostic methods are under development but are not yet available for general use. Serology can detect infection when parasites are too few to be identified in blood smears.

Treatment and Control

Chemotherapy is variably effective in treating infected birds or flocks. Chloroquine (5-10 mg/kg) potentiated with primaquine (0.3 mg/kg), or chloroquine in drinking water (250 mg/120 mL) has been used. Grape or orange juice can disguise chloroquine's bitterness. Other antimalarial drugs have been used with success; quinacrine at 1.6 mg/kg/day for 5 days was successful for treating an infected peacock. Treatment with a combination of sulfamonomethoxine and sulfachloropyrazine is effective; halofuginone can be used for chemoprophylaxis in endemic areas. Persistent parasitemia or relapse may occur during and after treatment. Treatments should be evaluated on prevention of mortality and improvement of clinical disease. Birds that survive initial infection are generally refractory to subsequent infections. Prevention of exposure to mosquitos is a useful adjunct. Parasite development is reduced in mosquitos that survive previous infection with Bacillus thuringiensis *israelensis*.

Other Bloodborne Organisms

Trypanosomes have been described in several avian species but rarely if ever cause clinical disease. They are more commonly identified in organ smears, especially bone marrow, than in peripheral blood, and can be cultured. Invertebrate hosts are thought to be any of several bloodsucking insects. Treatment is not warranted. Borreliae are tickborne (Argas spp) spirochetes that can cause fatal systemic disease. Tetracyclines, penicillin, and tick control are used for prevention and treatment. Small, punctate, or rarely ring-shaped, rickettsia-like basophilic bodies resembling Pirhemocyton, an organism that infects reptiles, are frequently seen in avian erythrocytes. Their identity and significance are unknown. They tend to be smaller and morphologically distinct from Aegyptianella.

Babesia spp are uncommon, nonpigmented, pyriform-shaped, erythrocytic protozoan parasites of birds. Natural infections of penguins, falcons, cranes, and several Asian avian species occur. Ticks are considered to be the invertebrate hosts. V, X, or fan shapes characterise dividing forms. Nothing is known of their significance, treatment, or control. Haemogregarina and Hepatozoon are protozoan parasites that are infrequently identified in birds. Both produce relatively large, nonpigmented, elongated gametocytes that can be found in RBC and WBC respectively. Gametocytes of Haemogregarina resemble gametocytes of Haemoproteus but lack pigment and are smaller, rarely if ever extending around or encircling the nucleus. Gametocytes of Hepatozoon are elongated with rounded ends and usually not located within an indentation of the nucleus, whereas Atoxoplasma is oval and partially encircled by the nucleus. Zoites of other sporozoa (eg, Toxoplasma, Sarcocystis) and organisms normally in the digestive tract (eg, trichomonads, coccidia, histomonads) may be transiently found in blood. The latter often also produce liver lesions.

8

Chicken Anaemia Virus Infection

Chicken Infectious Anaemia, Blue Wing Disease, Anaemia Dermatitis Syndrome, Hemorrhagic Aplastic Anaemia Syndrome

Chicken anaemia virus (CAV), a 25 nm, nonenveloped, icosahedral virus with a single-stranded, circular DNA genome, is the only member of the genus Gyrovirus of the Circoviridae family. The genome codes for 3 viral proteins (VP). VP1 is the capsid protein, but VP2 may be needed as a scaffold protein to allow proper folding of VP1. VP3, or apoptin, is a nonstructural protein that induces apoptosis in infected cells. CAV infects only chickens, although antibodies have been detected in Japanese quail. The virus is present worldwide based on serology and virus isolation. The disease, chicken infectious anaemia, has been described in most countries where chickens are raised commercially.

Horizontal transmission of CAV is by the faecal-oral route and perhaps by the respiratory route. Vertical transmission occurs when seronegative hens become infected and continues until neutralising antibodies develop. Chicks hatched from these eggs are viremic, and CAV can rapidly spread horizontally from these chicks to susceptible, maternal antibody-negative hatchmates. Roosters shedding CAV in the semen are another source of vertical transmission. Vaccination of seronegative flocks prior to the onset of egg production is recommended to prevent vertical transmission.

Maternal antibody-negative chicks are susceptible to infection and disease until 1-2 wk of age. In contrast, maternal antibody-positive chicks are protected from disease and probably from infection.

Age resistance to clinical disease, but not infection, begins at approximately 1 wk of age. The age resistance can be overcome by coinfection of CAV with immunosuppressive agents such as infectious bursal disease virus (*Infectious Bursal Disease: Introduction*), Marek's disease herpesvirus (*Marek's Disease*), and reticuloendotheliosis virus (*Reticuloendotheliosis*).

Many SPF flocks developed antibodies to CAV during or after the onset of sexual development. Spread of infection by CAV-contaminated embryo- or cell-culture-derived vaccines is possible.

When day-old susceptible chicks are inoculated IM with CAV, viraemia occurs within 24 hr. Virus can be recovered from most organs and rectal contents up to 35 days after inoculation. The principal sites of CAV replication are hemocytoblasts in the bone marrow, precursor T cells in the cortex of the thymus, and CD8 cells in the spleen. Replication in the first leads to anaemia, while replication in the latter two causes immunosuppression.

Neutralising antibodies are detectable 21 days after infection and clinical, hematologic, and pathologic parametres return to normal ~35 days after infection. CAV infection has adverse effects on proliferative responses of spleen lymphocytes and on the production of interleukin. and interferons by splenocytes. Infection can cause a marked decrease in generation of antigen-specific cytotoxic T cells directed against other pathogens.

In addition to T-cell defects, macrophage functions such as Fc-receptor expression, phagocytosis, and antimicrobial activity may be impaired. Subclinical, horizontally acquired infection with CAV in broiler progeny of seropositive parent flocks may be associated with impaired economic performance.

Clinical Findings

Signs of illness or adverse effects on egg production do not occur when seronegative adult chickens become infected. However, vertical transmission or infection of maternal antibody-negative chicks before 1 wk of age can cause clinical disease 12-17 days after hatching or infection. Chicks are anorectic, lethargic, depressed, and pale. PCV is low (in chicks, anaemia is defined as a PCV of d"27), and blood smears often reveal anaemia, leukopenia, or pancytopenia depending on the state of the disease. Blood may be watery and clot slowly. Mortality rates are variable but may be high with secondary complicating infections.

Lesions

Organs are pale; the thymus is generally atrophied, and the bursa of Fabricius may be small. Bone marrow is pale or yellow. Hemorrhage may be present in or under the skin, muscle, and other organs.

Histologically, lymphoid cell populations are depleted in primary and secondary lymphoid organs. Granulocytic and erythrocytic compartments in the bone marrow are atrophic or hypoplastic.

Diagnosis

A tentative diagnosis is based on history, signs, and gross and histopathologic lesions. Confirmation requires detection of virus or viral DNA in the thymus or bone marrow. PCR and quantitative PCR techniques are commonly used to demonstrate the presence of CAV.

Viral isolation can be used but is slow and expensive. To isolate CAV, chloroform-treated extracts of tissues are inoculated in MDCC-MSB1 or MDCC-147 cultures (a lymphoblastoid cell line derived from Marek's disease tumour) or into susceptible, immunocompromised (antigen- and antibody-negative), day-old chicks. Commercial ELISA kits are available to detect serum antibodies to CAV and can be used to identify breeder flocks that are seronegative prior to egg production and to monitor the efficacy of vaccination.

Treatment and Prevention

There is no specific treatment. Secondary bacterial infections may be treated with antibiotics. Live vaccines are available for vaccination of antibody-negative breeder flocks prior to the start of egg production. Administration is by injection or by addition to the drinking water depending on the type of vaccine available in individual countries.

In some areas, transfer of litter to noncontaminated premises and the addition of crude homogenates of tissues from affected chickens to the drinking water have been used to ensure infection and seroconversion of parent flocks before they begin to lay, thereby diminishing the risk of egg transmission.

However, these procedures are risky and not recommended. Because of the synergism between CAV and other immunosuppressive viruses, control of the latter is also important.

At present, there is no vaccine available to prevent subclinical losses in broilers.

Dissecting Aneurysm

Dissecting aneurysm is a fatal disease of turkeys characterised by sudden death of rapidly growing birds with massive internal hemorrhage resulting from rupture of aneurysms formed in various parts of the vascular system. The frequency with which the posterior aorta is affected has given rise to the term "aortic rupture." The disease has been reported in North America, Europe, and Israel. Most breeds of turkeys are susceptible, and the largest and most rapidly growing males, 8-24 wk old, are affected most often; females are also affected but at a lower incidence.

Etiology

The cause is unknown. Probably several factors contribute to the development of fatal cases. For the disease to occur, birds must be fed and managed in such a way that they are growing rapidly, and they must have a genetic susceptibility. A prolonged lipemia generally develops during the period of rapid growth, and the period of greatest mortality typically corresponds to a sharp rise in blood pressure, with dissecting aneurysms developing at the site of arteriosclerotic plaques. The lipemia may result from a high dietary intake of fat or from the effects of hormonal factors, such as high dietary concentrations of estrogens. Although â-aminopropionitrile, the toxic agent in Lathyrus odouratus, is capable of producing the disease, there is no evidence that this or other nitriles are responsible for dissecting aneurysms in turkeys under natural conditions. The enzyme lysyloxidase, isolated from turkey aortas and active on tropelaston and collagen cross-linking, was found to be much lower in males than females; this may be a factor in the development of spontaneous aortic aneurysms in male turkeys.

Clinical Findings

Affected birds that had shown no premonitory signs are found dead with marked pallor of the head and neck. Occasionally, a caretaker observes an apparently healthy bird die within a few minutes. The incidence is usually <1% but may be as great as 10%. Formerly, when male turkeys were implanted with stilbestrol, the incidence was as high as 20%.

Lesions

The carcass is markedly anaemic with large quantities of clotted blood in the peritoneal cavity and over the kidneys, or in the pericardial sac. The rupture in the ventral wall of the posterior aorta at about the position of the testes, or in the cardiac atrium, can be located readily by carefully washing away the blood clot. The aortic lumen may contain an organised, adherent thrombus at the site of rupture. Ruptures in smaller blood vessels are more difficult to locate.

Almost always, an intimal thickening or a large, fibrous plaque is present in the region of the rupture. The tunicas intima and media are thrown into deep folds and separated from the tunica adventitia. Marked accumulation of lipids in the thickened intima and in the fibrous plaques can be identified by stains. Fibres of the tunica media may show degenerative changes and infiltration with heterophils and macrophages.

Diagnosis

The diagnosis is made by finding large clots of blood in the coelomic cavity (aortic rupture) or within the pericardial sac (auricular rupture) of rapidly growing male turkeys. The condition should be differentiated from hypertensive angiopathy (*Perirenal Hemorrhage Syndrome of Turkeys: Introduction*), which is also seen in rapidly growing turkeys. In hypertensive angiopathy, the major lesions include pulmonary edema and supcapsular perirenal hemorrhage.

Treatment, Control, and Prevention

There is no known treatment. Coagulants and vitamin K are useless because there is no defect in the clotting mechanism. Losses sometimes may be reduced during the critical period between 16 and 23 wk of age by limiting feed intake or slowing growth rate by reducing the energy level of the diet. High-fat diets should not be fed during this period. Some studies have indicated that the incidence of aortic rupture can be reduced by adding copper at 125-250 ppm to the diet from at least 4 wk of age until market.

Inclusion Body Hepatitis Hydropericardium Syndrome

Hepatitis Hydropericardium

Adenoviruses are widespread throughout all avian species. Studies have demonstrated the presence of antibodies in healthy poultry, and viruses have been isolated from normal birds. Despite their widespread distribution, the majority of adenoviruses cause no or only mild disease; however, some are associated with specific clinical conditions. Avian adenoviruses (AAV) in chickens are the etiologic agents of 2 important diseases known as inclusion body hepatitis (IBH) and hydropericardium syndrome (HP). Although in some cases each condition is observed separately, during the last decade the 2 conditions have been frequently observed as a single entity; therefore, the name hepatitis hydropericardium has been widely used to describe the pathologic condition. The syndrome is an acute disease of young chickens associated with anaemia, hemorrhagic disorders, and hydropericardium. It is a common disease in several countries, where broilers are severely affected, resulting in high mortality rates.

Etiology, Transmission, and Pathogenesis

The AAV of group I are the etiologic agents of this condition. Although there are 12 different serotypes of AAV, the most common viruses isolated in cases of IBH/HP belong to serotypes 4 and 8. These AAV are capable of producing the disease without the immunosuppressive effects of associated viruses such as infectious bursal disease (IBDV, *Infectious Bursal Disease: Introduction*) or other immunosuppressive agents. However, the association with immunosuppressive viruses such as IBDV and chicken anaemia virus (CAV, *Chicken Anaemia Virus Infection: Introduction*) will result in a more severe disease.

Horizontal and vertical transmission play an important role in IBH/HP. Vertical transmission has been described in progeny from breeder flocks infected with AAV serotypes 4 and 8. Horizontal transmission has also been demonstrated; young chicks in contact with infected chicks can die of peracute IBH/HP.

Chicks and young chickens are commonly affected. Infection with some strains of AAV may result in minimal hepatic disease; however, if birds have been infected with immunosuppressive viruses (IBD, CAV, Marek's disease), the clinical disease becomes evident.

Clinical Findings, Lesions, and Diagnosis

Sudden mortality usually is seen in chickens <6 wk old and as young as 4 days of age. Mortality normally ranges from 2-40%, especially when birds are <3 wk of age. However, there have been outbreaks in which mortality has reached 80%. Mortality rates also vary depending on the pathogenicity of the virus and infection with other viral or bacterial agents. Signs associated with diseases caused by other pathogens (eg, bacteria, fungi, or viruses) commonly occur if birds are immunosuppressed.

Flocks of 3 to 5 wk old broilers with HP may not show specific clinical signs, but abrupt onset of mortality, lethargy, huddling with ruffled feathers, and yellow, mucoid droppings may be seen. The duration of the infection usually ranges from 9-14 days with morbidity of 10-30% and a daily mortality of 3-5%.

Gross lesions include up to 10 mL of a straw-coloured transudate in the pericardial sac, generalised congestion, and an enlarged, pale, friable liver. Histopathologic lesions include myocardial edema in the heart with degeneration, necrosis, and mild mononuclear cell infiltration.

Basophilic intranuclear inclusion bodies may be present in the liver. A tentative diagnosis is based on typical microscopic findings and confirmed by isolating adenoviruses from the liver. Serology, restriction enzyme analysis and PCR are used to classify adenoviruses isolated from clinical cases. This information is used for epidemiologic studies.

Treatment and Prevention

As with many other viral diseases, there is no treatment. Antibiotics may help prevent secondary bacterial infections. Sulfonamides are contraindicated if evidence of hematologic disease or immunosuppression is seen.

Vaccines against IBH/HP are not commercially available in the USA; however, in other countries both live and inactivated vaccines are used to control the syndrome. The AAV serotypes most frequently used to prepare commercial vaccines are serotypes 4 and 8. Primary breeders with stringent biosecurity practices sometimes use autogenous inactivated vaccines to ensure the transfer of maternal immunity from breeding flocks to their progeny.

In Australia, a live vaccine given via drinking water was developed for breeders between 10-14 wk of age. In other countries, including Mexico, Pakistan, and Peru, inactivated vaccines are routinely used to vaccinate breeders and broilers.

When breeders are properly vaccinated, antibodies generated by the vaccine are transmitted to the progeny, providing protection against field infections and clinical disease. Broilers are vaccinated at <10 days of age when their parents either do not have serotype-specific adenovirus antibodies or maternal antibody transmission is erratic due to improper vaccination procedures that result in a substantial number of unvaccinated birds.

Perirenal Hemorrhage Syndrome of Turkeys

Hypertensive Angiopathy, Sudden Death Syndrome of Turkeys

Perirenal hemorrhage syndrome (PHS) is a noninfectious cardiovascular disease usually affecting rapidly growing male turkeys 8-15 wk old characterised by sudden death, perirenal hemorrhage, and hypertrophic cardiomyopathy. Mortality is usually 0.5-2% but can be higher; there is no morbidity. Healthy, rapidly growing flocks are more likely to be affected.

The pathogenesis is unknown, but PHS is apparently unrelated to pulmonary function or hypertension. Inadequate or inappropriate cardiac response to exercise, resulting in hypotension, vasodilation, arrhythmias, and sudden death, appears most likely. Acute congestive heart failure secondary to cardiac hypertrophy has also been suggested as a potential cause. Renal hemorrhage may occur due to severe passive congestion.

Gross lesions include good to excellent body condition, food in crop and stomach, enlarged dark red to purple spleen, variable retroperitoneal hemorrhage around one or both kidneys, generalised congestion, and pulmonary edema occasionally accompanied by hemorrhage.

Cardiac hypertrophy involving the left ventricle and intraventricular septum may also be seen. Microscopic changes are consistent with gross findings; proliferative arterial and arteriolar lesions and ruptured renal veins also are often present. PHS has several characteristics in common with aortic rupture (*Dissecting Aneurysm: Introduction*) and flip-over in broilers (*Flip-over Disease: Introduction*).

Diagnosis is based on history, typical gross lesions, and absence of infectious agents. Extensive PHS lesions may resemble aortic rupture.

There is no specific treatment. Factors that decrease growth rate and activity also tend to decrease PHS. Reserpine (0.5 ppm feed) decreases PHS, but aspirin (0.005%) or increased calcium has no effect. Reserpine is not listed in the Feed Additive Compendium as approved for use in feed for turkeys. Increased room temperature and various lighting programs have also reduced PHS. Activities that increase cardiovascular stress (eg, moving birds, tilling litter, noise) should be minimised, especially between 7 and 15 wk of age. Lower ambient temperatures (55°F [13°C]), intermittent lighting, and leaving toes unclipped increase mortality from PHS. PHS may occur in healthy commercial male turkey flocks regardless of management practices used to prevent its occurrence.

Round Heart Disease of Turkeys

Spontaneous Cardiomyopathy

Spontaneous cardiomyopathy of young turkeys is characterised by sudden death due to cardiac arrest. It has been suggested that the condition should be called spontaneous cardiomyopathy to distinguish it from round heart disease of chickens, a different syndrome that is rarely recognised today.

The exact etiology of spontaneous cardiomyopathy in turkeys is unknown. However, studies using furazolidone to produce dilated cardiomyopathy in turkeys have indicated altered membrane transport resulting in myocardial failure. Creatine kinase, glycolysis, glycogen, myofibril, Krebs cycle enzymes, fatty acid oxidation, and soluble proteins are all reduced. The calcium-transport ATPase activity of the sarcoplasmic reticulum is increased. This pattern of biochemical changes is consistent with ischemia playing a role in the pathogenesis of spontaneous cardiomyopathy in turkeys.

While most deaths occur during the brooding period, the ratio of heart weight to body weight of affected birds is increased throughout the growing period. Market body weights of affected birds are reduced an average of 3 lb (1.4 kg). Some outbreaks of the condition have been associated with hypoxia during incubation of the eggs or during transportation of poults from the hatchery to the brood farm.

Most deaths from spontaneous cardiomyopathy occur during the first 4 wk of life, with mortality peaking at 2 wk. Many poults die suddenly, but some may have ruffled feathers, drooping wings, and a general unthrifty appearance. They may show laboured, gasping breathing before death. After 3 wk of age, mortality is sporadic. Characteristically, the affected poult in the first 4 wk of life has a greatly enlarged heart due to dilatation of both ventricles, congested lungs, and a swollen liver. Ascites, anasarca, pulmonary edema, and hydropericardium may or may not be present. In older poults, enlarged hearts are due to marked hypertrophy of the ventricles in addition to dilatation. Histologically, lesions of abnormal hearts are nonspecific and include congestion, damage of the myofibrils of the cardiocytes, and focal infiltration by lymphocytes.

Generally, diagnosis is based on history and gross findings at necropsy; although an ECG can be used, it is of little practical use. Sodium and polychlorinated biphenyls or related compounds may produce similar syndromes. No treatment is available. Good brooding practices may reduce mortality. Any toxins should be eliminated. Incubation, transportation, and early brooding ventilation conditions should be reviewed.

Candidiasis

Thrush, Crop Mycosis, Sour Crop

Improving sanitation and minimising antibiotic use in poultry help reduce the incidence of candidiasis. Affected birds can be treated with copper sulfate at 0.5 mg/L of drinking water, or 0.5 mg copper sulfate per kg of feed. Vinegar is used as a treatment for candidiasis at 15 mL/L of drinking water. Chlorhexidine is used for prevention or treatment at 2.5 mL/L of drinking water. Chlorine bleach at 0.1 mL/L of drinking water may help control the infection. All of these treatments lack FDA approval.

Coccidiosis

Coccidiosis is caused by protozoa of the phylum Apicomplexa, family Eimeriidae. In poultry, most species belong to the genus Eimeria and infect various sites in the intestine. The infectious process is rapid (4-7 days) and is characterised by parasite replication in host cells with extensive damage to the intestinal mucosa. Poultry coccidia are strictly host-specific, and the different species parasitise specific parts of the intestine. Coccidia are distributed worldwide in poultry and wild birds.

Etiology

Coccidia are almost universally present in poultry-raising operations, but clinical disease occurs only after ingestion of relatively large numbers of sporulated oocysts by susceptible birds. Both clinically infected and recovered birds shed oocysts in their droppings, which contaminate feed, dust, water, litter, and soil. Oocysts may be transmitted by mechanical carriers (eg, equipment, clothing, insects, and other animals). Fresh oocysts are not infective until they sporulate; under optimal conditions (70-90°F [21-32°C] with adequate moisture and oxygen), this requires 1-2 days. The prepatent period is 4-7 days. Sporulated oocysts may survive for long periods, depending on environmental factors. Oocysts are resistant to some disinfectants commonly used around livestock but are killed by freezing or high environmental temperatures. Pathogenicity is influenced by host genetics, nutritional factors, concurrent diseases, and species of the coccidium. Eimeria necatrix and E tenella are the most pathogenic in chickens because schizogony occurs in the lamina propria and crypts of Lieberkühn of the small intestine and ceca, respectively, and causes extensive hemorrhage. Most species develop in epithelial cells lining the villi. Protective immunity usually develops in response to moderate and continuing infection. True age-immunity does not occur, but older birds are usually more resistant than young birds because of earlier exposure to infection.

Clinical Findings

Signs range from decreased growth rate to a high percentage of visibly sick birds, severe diarrhea, and high mortality. Feed and water consumption are depressed. Weight loss, development of culls, decreased egg production, and increased mortality may accompany outbreaks. Mild infections of intestinal species, which would otherwise be classed as subclinical, may cause depigmentation. Survivors of severe infections recover in 10-14 days but may never recover lost performance.

Chickens

E tenella infections are found only in the ceca and can be recognised by accumulation of blood in the ceca and by bloody droppings. Cecal cores, which are accumulations of clotted blood, tissue debris, and oocysts, may be found in birds surviving the acute stage.

E necatrix produces major lesions in the anterior and middle portions of the small intestine. Small white spots, usually intermingled with rounded, bright- or dull-red spots of various sizes, can be seen on the serosal surface.

The white spots are diagnostic for E necatrix if clumps of large schizonts can be demonstrated microscopically. In severe cases, the intestinal wall is thickened, and the infected area dilated to 2-2.5 times the normal diametre.

The lumen may be filled with blood, mucus, and fluid. Fluid loss may result in marked dehydration. Although the damage is in the small intestine, the sexual phase of the life cycle is completed in the ceca. Oocysts of E necatrix are found only in the ceca. Due to concurrent infections, oocysts of other species may be found in the area of major lesions, misleading the diagnostician.

E acervulina, the most common infection, is characterised by numerous, whitish, oval or transverse patches in the upper half of the small intestine and may be easily distinguished on gross examination. The clinical course in a flock is usually protracted and results in poor growth, an increase in culls, and slightly increased mortality.

E brunetti is found in the lower small intestine, rectum, ceca, and cloaca. In moderate infections, the mucosa is pale and disrupted but lacking in discrete foci, and may be thickened. In severe infections, extensive coagulative necrosis and sloughing of the mucosa occurs throughout most of the small intestine.

E maxima develops in the small intestine, where it causes dilatation and thickening of the wall; petechial hemorrhage; and a reddish, orange, or pink viscous mucous exudate and fluid. The oocysts and gametocytes (particularly macrogametocytes), which are present in the lesions, are distinctly large.

E mitis is recognised as pathogenic in the lower small intestine. Lesions resemble moderate infections of E brunetti but can be distinguished by finding small, round oocysts associated with the lesion.

Turkeys

Only 4 of the 7 species of coccidia in turkeys are considered pathogenic— Eimeria adenoeides, E dispersa, E gallopavonis, and E meleagrimitis. E innocua, E meleagridis, and E subrotunda are considered nonpathogenic. Oocysts sporulate within 1-2 days after expulsion from the host; the prepatent period is 4-6 days.

E adenoeides and E gallopavonis infect the lower ileum, ceca, and rectum. The developmental stages are found in the epithelial cells of the villi and crypts. The affected portion of the intestine may be dilated and have a thickened wall. Thick, creamy material or caseous casts in the gut or excreta may contain enormous numbers of oocysts. E meleagrimitis chiefly infects the upper and mid small intestine. The lamina propria or deeper tissues may be parasitised, which may result in necrotic enteritis (*Necrotic Enteritis: Introduction*). E dispersa infects the upper small intestine and causes a creamy, mucoid enteritis that involves the entire intestine, including the ceca. Large numbers of gametocytes and oocysts are associated with the lesions.

Common signs in infected flocks include reduced feed consumption, rapid weight loss, droopiness, ruffled feathers, and severe diarrhea. Wet droppings with mucus are common. Clinical infections are seldom seen in poults >8 wk old. Morbidity and mortality may be high.

Ducks

A large number of specific coccidia have been reported in both wild and domestic ducks, but validity of some of the descriptions is questionable. Presence of Eimeria, Wenyonella, and Tyzzeria spp has been confirmed. T perniciosa is a known pathogen that balloons the entire small intestine with mucohemorrhagic or caseous material. Eimeria spp also have been described as pathogenic. Some species of coccidia of domestic ducks are considered relatively nonpathogenic. In wild ducks, infrequent but dramatic outbreaks of coccidiosis occur in ducklings 2-4 wk old; morbidity and mortality may be high.

Geese

The most striking coccidial infection of geese is that produced by Eimeria truncata, in which the kidneys are enlarged and studded with poorly circumscribed, yellowish white streaks and spots. The tubules are dilated with masses of oocysts and urates. Mortality may be high. At least 5 other Eimeria spp have been reported to parasitise the intestine.

Diagnosis

The location in the host, appearance of lesions, and the size of oocysts are used in determining the species present. Coccidial infections are readily confirmed by demonstration of oocysts in feces or intestinal scrapings; however, the number of oocysts present has little relationship to the extent of clinical disease.

Severity of lesions as well as knowledge of flock appearance, morbidity, mortality, feed intake, growth rate, and rate of lay are important for diagnosis. Necropsy of several fresh specimens is advisable. Classical lesions of E tenella and E necatrix are pathognomonic, but infections of other species are more difficult to diagnose. Comparison of lesions and other signs with diagnostic charts allows a reasonably accurate differentiation of the coccidial species. Mixed coccidial infections are common. A diagnosis of clinical coccidiosis is warranted if oocysts, merozoites, or schizonts are demonstrated microscopically and if lesions are severe. Subclinical coccidial infections may be unimportant, and poor performance may be caused by flock disorders.

Control

Practical methods of management cannot prevent infection. Poultry that are maintained at all times on wire floors to separate birds from droppings have fewer infections; clinical coccidiosis is seen only rarely under such circumstances. Other methods of control are vaccination or prevention with anticoccidial drugs.

Vaccination

A species-specific immunity develops after natural infection, the degree of which largely depends on the extent of infection and the number of reinfections. Protective immunity is primarily a T-cell response.

Commercial vaccines consist of low doses of live, sporulated oocysts of the various coccidial species administered at low doses to day-old chicks. Because the vaccine serves only to introduce infection, chickens are reinfected by progeny of the vaccine strain on the farm. The vaccine strains of coccidia may or may not be attenuated. The self-limiting nature of coccidiosis is used as a form of attenuation for some vaccines, rather than biological attenuation.

Layers and breeders that are maintained on floor litter must have protective immunity. Often, they are given a suboptimal dosage of an anticoccidial drug during early growth, with the expectation that immunity will continue to develop from repeated exposure to wild types of coccidia. This method has never been particularly successful because of the difficulty in controlling all of these factors. Immunity is not necessary in broiler chickens or cage layers. Prevention of infection by anticoccidial drugs is preferred.

Anticoccidial Drugs

Many products are available for prevention or treatment of coccidiosis in chickens and turkeys. Detailed instructions for use are provided by all manufacturers to help users comply with regulatory approvals and management considerations. High dosages may sometimes be used over short periods for treatment or if a high level of exposure is anticipated.

Anticoccidials are given in the feed to prevent disease and the economic loss often associated with subacute infection. Prophylactic use is preferred because most of the damage occurs before signs become apparent, and because drugs cannot completely stop an outbreak.

Water medication is generally preferred over feed medication for therapeutic treatment. Antibiotics and increased levels of vitamins A and K are sometimes used in the ration to improve rate of recovery and prevent secondary infections.

Continuous use of anticoccidial drugs promotes the emergence of drug-resistant strains of coccidia. Various programs are used in attempts to slow or stop selection of resistance. For instance, producers may use one anticoccidial continuously through succeeding flocks, rotate anticoccidials every 4-6 mo, or change anticoccidials during a single growout (ie, a shuttle program). While there is little cross-resistance to anticoccidials with different modes of action, there is widespread resistance to most drugs.

Change of drug may be beneficial when resistance has been established. "Shuttle programs," in which 1 group of chickens is treated sequentially with different drugs (usually a change between the starter and grower rations), are common practice in many countries, and offer some benefit in reducing emergence of resistance. In the USA, the FDA considers shuttle programs as extra-label usage, but producers may use such programs on the recommendation of a veterinarian.

The effects of anticoccidial drugs may be coccidiostatic, in which growth of intracellular coccidia is arrested but development may continue after drug withdrawal, or coccidiocidal, in which coccidia are killed during their development. Some anticoccidial drugs may be coccidiostatic when given short-term but coccidiocidal when given longterm. Most anticoccidials currently used in poultry production are coccidiocidal.

The natural development of immunity to coccidiosis can be slowed by use of some highly effective anticoccidials. In the production of broilers during a short growout of 37-44 days, this may be of little consequence. However, natural immunity is important in replacement layers because they are likely to be exposed to coccidial infections for extended periods after terminating anticoccidial drugs. Anticoccidial programs for layer and breeder flocks are aimed at allowing immunising infection while guarding against acute outbreaks.

Anticoccidials are commonly withdrawn from broilers 3-7 days before slaughter to meet regulatory requirements and to reduce production costs. Because broilers have varying susceptibility to infection at this point, the risk of coccidiosis outbreaks is increased with longer withdrawal.

Turkeys are given a preventive anticoccidial for confinement-reared birds up to 8-10 wk of age. Older birds are considered less susceptible to outbreaks. The modes of action of anticoccidial drugs are poorly understood. Some that are better known are described below. Knowledge of mode of action is important in understanding toxicity and side effects. Amprolium is structurally similar to and is a competitive antagonist of thiamine (vitamin B_1). Because rapidly dividing coccidia have a relatively high requirement for thiamine, amprolium has a safety margin of ~8:1 when used at the highest recommended level in feed. Maximal effect occurs about day 3 of the life cycle of coccidia. Because amprolium has poor activity against some Eimeria spp, its spectrum has been extended by using it in mixtures with the folic acid antagonists, ethopabate and sulfaquinoxaline.

Clopidol and quinolines (eg, decoquinate, methylbenzoquate) halt development of the sporozoites or trophozoites of Eimeria spp by inhibiting the electron transport system within parasite mitochondria. This action is coccidiostatic. Clopidol and quinolines have a broad species spectrum, but resistance may develop rapidly.

Folic acid antagonists include the sulfonamides, 2,4-diaminopyrimidines and ethopabate. These compounds are structural antagonists of folic acid or of para-aminobenzoic acid (PABA), which is a precursor of folic acid. (The host does not synthesize folic acid and has no requirement for PABA.) Coccidia rapidly synthesize nucleic acids, especially during schizogony, which accounts for activity against these stages. Although resistance to antifolate compounds is widespread, they are commonly used for water treatment when clinical signs are already evident. Diaveridine, ormetoprim, and pyrimethamine are active against the protozoan enzyme dihydrofolate reductase. They have synergistic activity with sulfonamides and often are used in mixtures with these compounds.

Halofuginone hydrobromide is related to the antimalarial drug febrifuginone and is effective against asexual stages of most species of Eimeria. It has both coccidiostatic and coccidiocidal effects.

The ionophores (monensin, salinomycin, lasalocid, narasin, maduramicin, and semduramicin) form complexes with various ions, principally sodium, potassium, and calcium, and transport these into and through biological membranes. The ionophores affect both extra- and intracellular stages of the parasite, especially during the early, asexual stages of parasite development. Drug tolerance was initially slow to emerge, probably because of the biochemically nonspecific way these fermentation products act on the parasite. Recent surveys suggest that drug tolerance is now widespread, but these products remain the most important class of anticoccidials.

Some ionophores depress weight gain when given at or slightly above the recommended levels. Primarily, this is the result of reduced feed consumption, but the reduced growth may be offset by improved feed conversion.

Nicarbazin was the first product to have truly broad-spectrum activity that is still in common use. While not completely understood, the mode of action is thought to be via inhibition of succinate-linked nicotinamide adenine dinucleotide reduction and the energy-dependent transhydrogenase, and the accumulation of calcium in the presence of ATP. Nicarbazin is toxic for layers, and a 4 day withdrawal period is required in broilers. Medicated birds are at increased risk of heat stress in hot weather.

Nitrobenzamides (eg, dinitolmide) exert their greatest coccidiostatic activity against the asexual stages. Efficacy is limited to E tenella and E necatrix unless combined with other products.

Robenidine, a guanidine compound, allows initial intracellular development of coccidia but prevents formation of mature schizonts. It is both coccidiostatic when given short term and coccidiocidal long term. Drug resistance may develop during use. A 5 day withdrawal period is needed to eliminate untoward flavour caused by residues in poultry meat.

Cryptosporidiosis

Cryptosporidiosis is caused by protozoa (phylum Apicomplexa) that are members of the family Cryptosporidiidae and are related to coccidia of the genera Eimeria, Isospora, Sarcocystis, and Toxoplasma. Until recently, it was thought that there were 19 species in the genus Cryptosporidium, but recent research has shown that most are merely species that lack host specificity. Cryptosporidia are parasitic in the intestine of mammals, but in birds they are commonly found in the bursa and in the respiratory tract. Cryptosporidiosis is more severe in turkeys than in chickens and is frequently fatal in quail.

The life cycle of Cryptosporidium is similar to that of other coccidia, involving asexual and sexual phases, and culminates in oocyst production.

In the host, the oocyst forms four sporozoites without sporocysts. The life cycle is not self-limiting (as with other coccidia) because some oocysts are thin-walled and release sporozoites (after trypsin/bile stimulation) that reinfect adjacent tissues. The endogenous cycle is short (4-7 days), the endogenous stages are small (4-7 µm), and the parasites are just beneath the epithelial cell membranes. In turkeys and chickens, Cryptosporidium have been found in the sinuses, trachea, bronchi, cloaca, and bursa. The respiratory disease causes coughing, gasping, and airsacculitis. Lungs become gray and wet. Signs last several weeks, and death may occur.

Examining tissue scrapings from the bursa, cloaca, and trachea, and finding the characteristic small (5 µm) oocysts can be diagnostic. Concentration of intestinal scrapings using saturated sugar solution and examination by phase-contrast or interference-contrast microscopy is preferred.

There are no satisfactory control measures except isolation and good sanitation. All known anticoccidial drugs are ineffective against Cryptosporidium spp. Unlike Cryptosporidium spp of other mammals, the avian species are not infectious to people.

Coronaviral Enteritis of Turkeys

Etiology and Epidemiology

Coronaviral enteritis is an acute, highly contagious disease of turkeys characterised by sudden onset, marked depression, anorexia, diarrhea, dehydration, and weight loss. Mortality may be high, particularly in poults, but failure to gain body weight in adult birds may be more important economically.

Etiology and Epidemiology

The causative agent is a coronavirus, but the clinical disease is often complicated by other intestinal viral, bacterial, and protozoal infections. Spread is by direct or indirect contact with infected birds or contaminated premises. Droppings of acutely infected birds are rich in virus, and recovered birds may continue to shed lower levels of virus for months. Environmental factors do not appear to influence the occurrence; however, cold temperatures may contribute to the severity of the disease. Cold weather, especially freezing, and high litter moisture increase survival of the virus.

Clinical Findings

A short incubation period, often 48-72 hr, is followed by general depression, anorexia, and diarrhea in the flock. Young poults appear cold, chirp constantly, and seek heat. Feed and water consumption drop markedly, and poults lose weight rapidly. Morbidity and mortality may approach 100% in uncontrolled outbreaks. Morbidity and mortality are variable in growing and adult turkeys. Profuse diarrhea, with mucoid threads or casts in the droppings, is common. Dehydration and weight loss are often pronounced, and several weeks may be required for normal growth to resume. Cyanosis of the head is common. Breeder hens experience a severe drop in egg production and produce abnormal eggs with chalky shells. Vertical transmission does not occur.

Lesions

Young birds have few lesions other than flaccid, distended intestines that contain excess fluid and gas. Ceca are distended with foamy, pale brown, fetid fluid. Lesions in older birds are more extensive. Skin and musculature are dehydrated, and petechial hemorrhages may be seen on the viscera. Kidneys frequently are swollen and contain an excess of urates, and the pancreas may have multiple, chalky white areas. Severe catarrhal enteritis is common and mucoid casts may be present. The crop may be distended and contain sour-smelling fluid. The spleen is often small and pale gray.

Diagnosis

Although clinical findings and lesions are suggestive, definitive diagnosis requires laboratory techniques including demonstration of coronaviral antigen in intestines of affected birds by direct fluorescent antibody techniques, detection of coronavirus particles in intestinal contents by electron microscopy, reproduction of the disease in naive poults with bacteria-free intestinal filtrates, and negative findings for common bacterial and protozoal infections. Other conditions that may produce similar signs in poults include hexamitiasis (*Hexamitiasis: Introduction*), salmonellosis (*Salmonelloses: Introduction*), inanition, and water deprivation. Other intestinal viruses (which are common in commercial flocks, including rotavirus, reovirus, astrovirus, enterovirus, and possibly others) can cause disease that resembles mild coronaviral enteritis. In older birds, severe larval ascarid infection may cause diagnostic confusion.

Prevention and Treatment

Introduction of virus should be minimised by good management and sanitation practices. Depopulation of problem premises followed by thorough cleaning and disinfection of buildings and equipment is effective in breaking the cycle of infection. Such farms are best cleaned during summer and should be left vacant for e"1 mo.

A commercial vaccine is not available. "Controlled" exposure programs have been used with variable success on some problem farms, but such procedures are not recommended because carrier states may be induced.

The course of disease outbreaks may be altered by good nursing care and judicious use of antibiotics and other drugs to combat secondary bacterial infections and dehydration. Birds in brooder houses should be provided with supplemental heat, and birds on range should be protected from adverse environmental conditions. Antibiotic administration decreases mortality but not the growth suppressant effects. The selection of an antibiotic is empiric at best, but tetracyclines, neomycin, streptomycin, lincomycin, penicillin, and bacitracin are among those used with variable success. Antibiotics may be added to drinking water in combination with calf milk-replacer and electrolyte, eg, 25 lb (11.4 kg) of calf milk-replacer and 450 g of potassium chloride to 100 gal. (380 L) of water. Birds should be medicated for 7-10 days. During and after treatment, birds should be observed closely for secondary crop mycosis, a common sequela of longterm antibiotic therapy.

9

Duck Viral Enteritis

Duck Plague

Duck viral enteritis (DVE) is an acute, highly contagious disease of ducks, geese, and swans of all ages, characterised by sudden death, high mortality (particularly among older ducks), and hemorrhages and necrosis in internal organs. It has been reported in domestic and wild waterfowl in Europe, Asia, North America, and Africa, resulting in limited to serious economic losses on domestic duck farms and sporadic, limited to massive die-offs in wild waterfowl.

Etiology, Epidemiology, and Transmission

Field strains of the causative herpesvirus are antigenically similar but vary considerably in pathogenicity. The virus is relatively sensitive to heat and pH; lipid solvents, trypsin, and chymotrypsin inactivate it. It causes intranuclear inclusion bodies in infected tissues and in inoculated cell cultures.

In nature, the virus is mainly transmitted from infected to susceptible ducks by direct contact or water and is acquired mainly by the oral route. Parenteral, intranasal, or oral administration of infected tissues can establish experimental infection. Recovered birds may remain carriers, serving as uncontrolled sources of the virus for susceptible ducks.

Clinical Findings

The incubation period is 3-7 days. Sudden high and persistent mortality is often the first sign of the disease. Mortality varies from 5-100% depending on the virulence of the infecting viral strain. Adult ducks usually die in higher proportions than young ones, increasing the economic significance of the disease. Dead males may have prolapse of the penis. Photophobia, inappetence, extreme thirst, droopiness, ataxia, nasal discharge, soiled vents, and watery or bloody diarrhea may be seen. Adult ducks may die in good flesh. In contrast, ducklings frequently show dehydration and weight loss, as well as blue beaks and bloodstained vents. In laying flocks, egg production may drop sharply.

Lesions

Hemorrhages in various tissues and free blood in body cavities indicate severe damage to blood vessels throughout the body. Petechial and ecchymotic hemorrhages on the heart ("paint brush" appearance), liver, pancreas, mesentery, and other organs are characteristic. Specific mucosal eruptions, found in the oral cavity, esophagus, ceca, rectum, and cloaca, undergo progressive alterations during the course of the disease.

Macular hemorrhages initially develop into elevated, yellowish, crusted plaques and organise into green, superficial scabs, which may coalesce into large, patchy, diphtheritic membranes. The mucosal lesions align parallel with the longitudinal folds in the esophagus and with the annular bands in the intestines. All lymphoid organs are affected; necrosis and hemorrhages are apparent. A lesion that can be easily detected on necropsy is a clear, yellow fluid that infiltrates and discolours the subcutaneous tissues from the thoracic inlet to the upper third of the neck. Ruptured yolk and free blood may be found in the abdominal cavity of laying ducks.

Diagnosis

Presumptive diagnosis is based on disease history and lesions. Definitive diagnosis requires laboratory work. Isolation of the virus from liver, spleen, or kidney tissues may be attempted in various cell cultures (preferably primary Muscovy duck embryo fibroblasts or Muscovy duck embryo liver cultures), duck embryos, or ducklings.

Inoculating the chorioallantoic membrane of 9 to 14 day old embryonated Muscovy duck eggs may result in isolation of the virus, but this method is not as sensitive as intramuscular inoculation of day-old ducklings. Muscovy ducklings are more susceptible than White Pekin ducklings. Neutralisation with specific antiserum in these systems confirms the identity of the virus. Fluorescent antibody test can demonstrate DVE viral proteins, and PCR, using DVE virus-specific primers, can amplify the viral DNA in duck tissues or inoculated cultured cells. Serologic tests have little value in the diagnosis of acute infections.

Differential diagnoses include duck viral hepatitis, pasteurellosis, necrotic and hemorrhagic enteritis, trauma, drake damage, and various toxicoses. Newcastle disease, avian influenza, and fowlpox may cause similar lesions but are rarely reported in ducks. Established cases should be reported to the appropriate regulatory agency.

Prevention, Treatment, and Control

There is no treatment. Contact with wild, free-flying waterfowl and direct or indirect contact with contaminated birds or material (free-flowing water) should be avoided. Control is effected by depopulation, removal of birds from the infected environment, sanitation, and disinfection.

Prevention is based on maintaining susceptible birds in a disease-free environment or immunisation.

A chicken-embryo-adapted, modified live virus vaccine has been approved for use in domestic ducks, in zoological aviaries, and by private aviculturists.

A 0.5 mL dose is administered subcutaneously or intramuscularly to domestic ducklings >2 wk of age with a booster inoculation a year later. The vaccine is not approved for use in wild ducks. An inactivated vaccine, which appears to be efficacious as the modified live vaccine, has not been tested on a large scale and is not currently licensed.

Hexamitiasis

Hexamitiasis is an acute, catarrhal enteritis of turkeys, pheasants, quail, chukar partridges, and peafowl. The highest mortality occurs in birds 1-9 wk old. Natural infection has not been observed in chickens. Pigeons are susceptible to another species of Hexamita (H columbae). Hexamitiasis is rare in North America.

Etiology

The causative protozoan parasite in turkeys, Hexamita meleagridis, is spindle-shaped, averages 8 × 3 μm, and has 6 anterior and 2 posterior flagella. It has not yet been cultured in experimental media, although it has been grown in the allantoic cavity of developing chicken and turkey embryos. It is transmitted directly by ingestion of contaminated feces. Encysted hexamitids may be more important in transmission than free flagellates. Many survivors become carriers and shed parasites in their droppings.

Clinical Findings and Lesions

The nonspecific signs include watery diarrhea, dry unkempt feathers, listlessness, and rapid weight loss despite the fact that the birds continue to eat. Birds may die in convulsions. Bulbous dilatations of the small intestine (especially duodenum and upper jejunum) filled with watery contents are characteristic. The crypts of Lieberkühn contain myriad H meleagridis, which attach to the epithelial cells by their posterior flagella.

Diagnosis

Diagnosis depends on finding the flagellates by microscopic examination of scrapings of the duodenal and jejunal mucosa. Hexamita spp move with a rapid, darting motion (in contrast to the jerky motion of trichomonads). To avoid contamination of instruments with other cecal protozoa, the duodenum should be opened first. Hexamita spp may be demonstrated in poults that have been dead for several hours if the scrapings are placed in a drop of warm (104°F [40°C]), isotonic saline solution on the slide. Presence of a few Hexamita in birds >10 wk old may be unimportant.

Prevention and Treatment

Because many birds remain carriers, breeder turkeys and poults should be raised on separate premises if possible, preferably with separate attendants. Wire platforms should be used under feeders and waterers. Pheasants and quail may also be carriers. There is no effective treatment for hexamitiasis, although oxytetracycline (0.22% in the feed for 2 wk) or chlortetracycline (0.022-0.044% in the feed for 2 wk) may be of some benefit.

Necrotic Enteritis

Necrotic enteritis is an acute enterotoxemia. The clinical illness is usually very short and often the only signs are a sudden increase in mortality. The disease primarily affects broiler chickens (2-5 wk old) and turkeys (7-12 wk old) raised on litter but can also affect commercial layer pullets raised in cages.

Etiology and Pathogenesis

The causative agent is the gram-positive, obligate, anaerobic bacteria Clostridium perfringens. It is usually isolated on blood agar, incubated anaerobically at 37°C, on which it produces a double zone of hemolysis. There are 2 primary C perfringens types, A and C, associated with necrotic enteritis in poultry. Toxins produced by the bacteria cause damage to the small intestine, liver lesions, and mortality.

C perfringens is a nearly ubiquitous bacteria readily found in soil, dust, feces, feed, and used poultry litter. It is also a normal inhabitant of the intestines of healthy chickens. The enterotoxemia that results in clinical disease most often occurs either following an alteration in the intestinal microflora or from a condition that results in damage to the intestinal mucosa (eg, coccidiosis, mycotoxicosis, salmonellosis, ascarid larvae). High dietary levels of animal byproducts (eg, fishmeal), wheat, barley, oats, or rye predispose birds to the disease. Anything that promotes excessive bacterial growth and toxin production or slows feed passage rate in the small intestine could promote the occurrence of necrotic enteritis.

Clinical Findings and Lesions

Most often the only sign of necrotic enteritis in a flock is a sudden increase in mortality. However, birds with depression, ruffled feathers, and diarrhea may also be seen. The gross lesions are primarily found in the small intestine (jejunum), which may be ballooned, friable, and contain a foul-smelling, brown fluid. The mucosa is usually covered with a tan to yellow pseudomembrane often referred to as a "Turkish towel" in appearance. This pseudomembrane may extend throughout the small intestine or be only in a localised area. The disease persists in a flock for 5-10 days, and mortality is 2-50%.

Diagnosis

A presumptive diagnosis is based on gross lesions and a gram-stained smear of a mucosal scraping that exhibits large, gram-positive rods. Histologic findings consist of coagulative necrosis of one-third to one-half the thickness of the intestinal mucosa and masses of short, thick bacterial rods in the fibrinonecrotic debris. Isolation of large numbers of C perfringens, from intestinal contents that produce the double zone of hemolysis as described above, can confirm the diagnosis. Double zone hemolysis should not be used as the sole criteria for identification of C perfringens because some strains do not produce both toxins responsible for the hemolysis characteristics. Differential media specifically designed for isolation of C perfringens is available and may be useful for diagnosis.

Necrotic enteritis must be differentiated from lesions produced by Eimeria brunetti and also from ulcerative enteritis. Uncomplicated coccidiosis rarely produces lesions as acute or severe as those seen with necrotic enteritis. Ulcerative enteritis caused by C colinum usually produces focal lesions from the distal portion of the small intestine (ileum) to the ceca and is almost always accompanied by hepatic necrosis.

Prevention, Control, and Treatment

Because C perfringens is nearly ubiquitous, it is important to prevent changes in the intestinal microflora that would promote its growth. This can be accomplished by adding antibiotics in the feed such as virginiamycin (20 g/ton feed), bacitracin (50 g/ton feed), and lincomycin (2 g/ton feed).

The addition of anticoccidial compounds, especially of the ionophore class, has been extremely helpful in preventing the coccidial damage that leads to necrotic enteritis. Avoiding drastic changes in feed and minimising the level of fishmeal, wheat, barley, or rye in the diet can also aid in the prevention of necrotic enteritis. Administration of probiotics or competitive exclusion cultures has been used to both prevent and treat clinical necrotic enteritis (presumably by preventing the proliferation of C perfringens). Treatment for necrotic enteritis is most commonly administered in the drinking water, with bacitracin (200-400 mg/gal. for 5-7 days), penicillin (1,500,000 u/gal. for 5 days), and lincomycin (64 mg/gal. for 7 days) most often used. In each case, the medicated drinking water should be the sole source of water. Moribund birds should be removed promptly, as they can serve as a source of toxicosis or infection due to cannibalism.

Rotaviral Infections in Chickens, Turkeys, and Pheasants

Rotaviral infections are characterised by enteritis and diarrhea in young birds, but chickens have been infected without showing clinical signs.

Avian rotaviruses consist of 4 distinct serotypes (A-D). Group A rotaviruses share a common group antigen with mammalian rotaviruses. Group D rotaviruses have been identified only in avian species. The relationships of the other 2 avian serotypes to mammalian serotypes have not been established. Transmission is horizontally by the oral route. Egg transmission has not been reported. Early signs of diarrhea (wet litter), depression, and poor or abnormal appetite can be seen 2-5 days after infection. Dehydration occurs rapidly, and mortality can be as high as 30-50% in pheasants and turkeys. The survivors appear healthy but smaller than normal. Lesions consist of dilated intestines filled with yellowish, watery contents with gas bubbles. Often, the carcass is dehydrated. Mortality is variable and is usually due to dehydration and emaciation.

Early diarrhea and inappetence that sometimes end with death are indicative but not pathognomonic of rotaviral infection. faecal samples or intestinal contents can be examined by electron microscopy with negative staining, either directly or after ultracentrifugation. Numerous rotaviral particles ~70 nm in diametre, with double-shelled capsids, can be seen and are distinguishable from reovirus by their more sharply defined outer edges.

For viral isolation in chicken-embryo liver cells or chick kidney cells, faecal material must be treated with trypsin. Isolated rotaviruses belong mostly to serotype A and, in general, do not cause cytopathic effects on primary isolation. The presence of virus can be demonstrated 2-3 days after inoculation by immunofluorescent staining.

No commercial vaccines are available. Thorough cleaning and disinfection of infected houses is advisable to limit infection. There is no specific treatment.

Trichomoniasis

Trichomoniasis in domestic fowl, pigeons, doves, and hawks is characterised, in most cases, by caseous accumulations in the throat and usually by weight loss. It has been termed "canker," "roup," and, in hawks, "frounce."

Etiology

The causative organism is Trichomonas gallinae, a flagellated protozoan that lives in the sinuses, mouth, throat, esophagus, and other organs. It is more prevalent among domestic pigeons and wild doves than among domestic fowl, although severe outbreaks have been reported in chickens and turkeys. Some strains of T gallinae cause high mortality in pigeons and doves. Hawks may become diseased after eating infected birds and commonly show liver lesions, with or without throat involvement. Pigeons and doves transmit the infection to their offspring in contaminated pigeon milk. Contaminated water is probably the most important source of infection for chickens and turkeys.

Clinical Findings

The disease course is rapid. The first lesions appear as small, yellowish areas on the oral mucosa. They grow rapidly and coalesce to form masses that frequently completely block the esophagus and may prevent the bird from closing its mouth. Much fluid may accumulate in the mouth.

There is a watery ocular discharge and, in more advanced stages, exudate about the eyes that may result in blindness. Birds lose weight rapidly, become weak and listless, and sometimes die within 8-10 days.

In chronic infections, birds appear healthy, although trichomonads can usually be demonstrated in scrapings from the mucous membranes of the throat.

Lesions

The bird may be riddled with caseous, necrotic foci. The mouth and esophagus contain a mass of necrotic material that may extend into the skull and sometimes through the surrounding tissues of the neck to involve the skin.

In the esophagus and crop, the lesions may be yellow, rounded, raised areas, with a central conical caseous spur, often referred to as "yellow buttons." The crop may be covered by a yellowish, diphtheritic membrane that may extend to the proventriculus.

The gizzard and intestine are not involved. Lesions of internal organs are most frequent in the liver; they vary from a few small, yellow areas of necrosis to almost complete replacement of liver tissue by caseous necrotic debris. Adhesions and involvement of other internal organs appear to be contact extensions of the liver lesions.

Diagnosis

Lesions of T gallinae infection are characteristic but not pathognomonic; those of pox and other infections can be similar. Diagnosis should be confirmed by microscopic examination of a smear of mucus or fluid from the throat to demonstrate the presence of trichomonads.

Trichomonads can be cultured easily in various artificial media such as 0.2% Loeffler's dried blood serum in Ringer's solution or a 2% solution of pigeon serum in isotonic salt solution. Good growth is obtained at 98.6°F (37°C). Antibiotics may be used to reduce bacterial contamination.

Control

Because T gallinae infection in pigeons is so readily transmitted from parent to offspring in the normal feeding process, chronically infected birds should be separated from breeding birds. In pigeons, recovery from infection with a less virulent strain of T gallinae appears to provide some protection against subsequent attack by a more virulent strain. Successful treatments include metronidazole (60 mg/kg body wt) and dimetridazole (50 mg/kg body wt, PO; or in the drinking water at 0.05% for 5-6 days). Neither of these drugs is approved for use in birds in the USA.

Ulcerative Enteritis

Ulcerative enteritis was first diagnosed in bobwhite quail (Colinus virginianus). It also affects chickens, turkeys, pheasants, grouse, and other gallinaceous birds.

The disease has also been reported in pigeons. Japanese quail (Coturnix coturnix *japonica)* are resistant, as only experimentally induced cases were reported in highly inbred populations. Marked differences in mortality between males and females suggest that susceptibility is an inheritable trait in *Coturnix* quail. Ulcerative enteritis occurs worldwide and may be acute or chronic.

Etiology

Clostridium colinum is the etiologic agent. It is an anaerobic, fastidious to culture, gram-positive, spore-forming, slightly curved rod, ~1 × 3-4 μm wide, with subterminal, oval spores. In chickens, the disease is a complex that is linked to stress, coccidiosis, infectious bursal disease, and other predisposing factors. To induce experimental disease in quail, >10 viable bacterial cells must be administered PO; chickens inoculated at the same levels are not affected.

Epidemiology

Birds that develop chronic ulcerative enteritis or have recovered from the disease remain carriers. Infection can be introduced by flies feeding on contaminated faecal material or by recovered carrier birds. Infected birds shed the bacterium in their droppings. Bobwhite quail are the most susceptible to this highly contagious disease. Most cases are reported in captive populations of bobwhite quail, suggesting that management plays a role in the incidence of ulcerative enteritis. C colinum spores can survive in the premises for months.

Pathogenesis

After oral infection, the bacterium adheres to the intestinal villi, producing enteritis and ulcers in portions of the small intestine and upper large intestine. Bacilli migrate to the liver via portal circulation, producing necrotic foci that later coalesce into extensive hepatic necrosis.

Infarcts of the spleen are common. Stained smears of the lesions reveal the rod-shaped C colinum microorganism. Although toxigenicity tests in mice have been negative, the role of an in situ-produced toxin in the pathogenesis has been suggested but not demonstrated.

Clinical Findings

In susceptible bobwhite quail, sudden death occurs without signs or weight loss and with up to 100% mortality in just 2-3 days. Acute lesions include hemorrhagic enteritis of the duodenum. In chickens, as well as other game birds, the course of the disease is less severe and is accompanied by anorexia.

Signs are similar to those seen in coccidiosis—depressed, listless birds with humped backs, ruffled feathers, diarrhea, sometimes bloody or watery white droppings, especially in quail in the prolonged course. Chickens recover within 2-3 wk and mortality rarely exceeds 10%.

Lesions

In early disease stages, the most common lesions include small, round ulcers surrounded by hemorrhages in the small intestine, ceca, and upper large intestine. Small ulcers later coalesce to form larger, sometimes perforating ulcers, producing local or diffuse peritonitis. The presence of blood in the gut resembles coccidiosis. Characteristic yellow to gray necrotic foci are the predominant lesions in the hepatic parenchyma. Spleen enlargement with hermorrhages and nectrotic areas may be present.

Diagnosis

Gross postmortem lesions including intestinal ulcerations and yellow to gray necrotising lesions in the liver assist in diagnosis. C colinum can be seen in gram-stained smears of the liver and intestinal lesions. In bacteremic birds, the microorganism can also be found in blood and spleen smears. In chickens, differentiating ulcerative enteritis from coccidiosis (*Coccidiosis: Introduction*) may be difficult as both diseases may be present simultaneously. Necrotic enteritis (*Necrotic Enteritis: Introduction*) and histomoniasis (*Histomoniasis: Introduction*) may also present a diagnostic problem, but the hepatic lesions of ulcerative enteritis help differentiate it from these diseases. C colinum can be isolated from liver samples cultured in strict anaerobic conditions in prereduced blood glucose-yeast horse plasma medium. A fluorescent antibody test also has been used to accurately diagnose ulcerative enteritis.

Prevention, Treatment, and Control

Bacitracin in the feed at 200 g/ton is used for prevention in quail. Streptomycin (0.006%) and furazolidone (0.02%s) in the feed are effective for treating the disease. Prevention must start with good management practices (eg, avoiding the introduction of new birds into existing flocks. High population density is a predisposing factor. The use of cages is recommended in quail breeding. Sick or dead birds should be removed promptly. Total cleanup between flocks and pest control in and around the premises are good preventive measures.

Avian Campylobacter Infection

Campylobacteriosis is a significant enterocolitis of humans acquired through consumption of undercooked poultry meat contaminated with Campylobacter jejuni.

This organism colonises the intestine of chickens, turkeys, and waterfowl but is generally nonpathogenic in mature poultry. Some strains of C jejuni can cause enteritis and death in newly hatched chicks and poults; however, it has not been possible to satisfy Koch's postulates and reproduce the syndrome previously termed "avian vibrionic hepatitis" by administering isolates of C jejuni to chickens. Commercial poultry and free-living birds are natural reservoirs of the thermophilic campylobacters (C jejuni, C coli, and C lari) and other poorly defined species. It is estimated that over half of all commercial broiler and turkey flocks harbour C jejuni. The organism has been isolated from numerous birds, including Columbae and domestic and free-living Galliformes and Anseriformes. C jejuni has been demonstrated in all areas of commercial poultry production. Isolation of the organism is a function of surveillance and ability of laboratory personnel to culture and identify Campylobacter spp.

Etiology and Epidemiology

Campylobacter jejuni is the predominant species associated with foodborne infection derived from poultry. Campylobacter coli and C lari are occasionally recovered from the intestinal tract of poultry and have also been implicated in foodborne infection. Environmental contamination is the source of infection for poults, chicks, and ducklings. Litter can remain infective for long periods, subject to at least a 10% moisture level and neutral pH. Infected chicks and poults become colonised and can continue to excrete C jejuni for their lifetimes. Contaminated water may introduce infection into poultry flocks, and nonchlorinated water derived from a dam, river, or shallow well should be regarded as a possible source.

Rats, wild birds, and houseflies can infect flocks; equipment and footwear contaminated with feces from an infected source may also serve as a vehicle of transmission. Once C jejuni has been introduced into the environment, rapid transmission within the flock occurs, with subsequent colonisation of a high proportion of exposed breeders, commercial-meat, or laying-strain poultry. It is unclear whether C jejuni can be transmitted vertically, either on the surface of eggs or by transovarial transmission. It can be isolated from the reproductive tracts of hens and roosters.

Clinical Findings

Many chicks are colonised with Campylobacter spp early in life with no associated clinical signs or pathology. Highly pathogenic isolates derived from people with enterocolitis may induce some mortality in chicks.

Lesions

Gross lesions in challenged chicks include distention of the jejunum, disseminated hemorrhagic enteritis, and in some cases, focal hepatic necrosis. Microscopic lesions include edema of the mucosa of the ileum and cecum with C jejuni in the brush border of enterocytes. Mononuclear infiltration of the submucosa and villous atrophy occur, with intraluminal accumulation of mucus, erythrocytes, and both mononuclear and polymorphonuclear cells. It is unclear whether these findings represent a true clinical syndrome in chicks.

Diagnosis

faecal specimens should be collected using rayon-tipped swabs, then placed in semisolid Cary-Blair transport medium. Enrichment culture of specimens in semisolid motility medium facilitates isolation when small numbers of C jejuni are present in a sample. Campylobacter should be cultured on selective media containing brucella agar base and bovine blood with up to 7 antibiotics that inhibit overgrowth of other Enterobacteriaceae. Thermophilic Campylobacter spp should be cultured at 42°C under microaerophilic conditions for 48 hr. The microaerophilic conditions generally consist of 85% nitrogen, 10% carbon dioxide, and 5% oxygen; however, some strains require a hydrogen-enriched atmosphere (5%). Campylobacter spp of significance in poultry are oxidase- and catalase-positive, indole-negative, and reduce selenite. The thermophilic species may be characterised on the basis of hippurate hydrolysis; nalidixic acid sensitivity is no longer reliable due to the increasing prevalence of fluoroquinolone resistant C jejuni. The Penner or Lior serotyping schemes can be used to classify C jejuni ribotyping, or pulsed-field gel analysis can distinguish among various C jejuni isolates.

Control and Prevention

Because C jejuni does not occur as a specific pathogen under commercial conditions, treatment of poultry flocks is not a consideration. If C jejuni is considered a problem in companion bird aviaries or in exotic species, antibiotics such as erythromycin can be administered in drinking water. Galliformes should receive a dosage of 10-30 mg/kg for 4 consecutive days, and Psittaciformes and exotics should be medicated at 30-40 mg/kg.

Preharvest prevention of Campylobacter infection in commercial species is based on strict biosecurity, decontamination of housing between successive flocks, exclusion of rodents and wild birds, and insect eradication. Chlorination of drinking water to 2 ppm and operation of farms on a strict "all-in/all-out" basis occasionally reduces the prevalence of infection. In the context of commercial production in the USA where earth-floored housing is used and litter is recycled, preharvest control of C jejuni is impractical. Innovative methods of prevention, such as competitive inhibition or the use of vaccines, are under intensive investigation, but are unlikely to be available for commercial application in the near future. Withholding feed from broilers and turkeys for at least 12 hr before slaughter and thorough decontamination of transport coops and modules reduce faecal contamination and lower the level of C jejuni introduced into processing plants.

Zoonotic Risk

C jejuni is a major source of foodborne enteritis in consumers; contaminated, undercooked poultry is responsible for >50% of cases investigated. The condition was recognised in the mid 1970s, and the significance of the organism has become apparent with improved methods of isolation and identification. Nonchlorinated ground water, unpasteurised milk, young diarrheic pets, and contaminated beef and pork products may also be responsible for infection of people.

Improved washing of carcasses, use of counter-flow scalding, elimination of immersion chillers, and reduction in manual handling by installation of advanced automated equipment can reduce C jejuni contamination. Chemical disinfectants, such as glutaraldehyde (0.125%) and succinic acid (3%), and organic compounds, such as lactic and acetic acids, may be used to destroy C jejuni.

Gamma irradiation at levels of 1-3 kGy effectively eliminates C jejuni from poultry carcasses and products. Irradiation using cobalt 60 and electron beam generation are cost-effective procedures, which are endorsed by a joint United Nations Committee of the Food and Agricultural Organisation, the International Atomic Energy Agency, and the World Health Organisation. However, irradiation is not well accepted by American consumers. Currently, the only measure to reduce the risk of C jejuni infection to consumers is thorough cooking of poultry to achieve a core temperature of 74°C for 1 min. This ensures destruction of C jejuni. Concurrent hygienic storage, handling, and preparation are necessary to prevent contamination of prepared foods, work surfaces, and utensils by raw poultry and other meats.

Avian Chlamydiosis

Psittacosis, Ornithosis, Parrot Fever

Avian chlamydiosis can be an inapparent subclinical infection or acute, subacute, or chronic disease of wild and domestic birds characterised by respiratory, digestive, or systemic infection. Infections occur worldwide and have been identified in at least 150 avian species, particularly colonial nesting birds (eg, egrets, herons), ratites, caged birds (primarily psittacines), raptors, and poultry. Among poultry, turkeys, ducks, and pigeons are most often affected; infection of chickens is infrequent.

The disease is a significant cause of economic loss and human exposure in European duck flocks. Longterm inapparent infections lasting for months to years are common and considered the normal chlamydia-host relationship; 10-30% of surveyed avian populations may be found positive. The same strain may cause mild disease or asymptomatic infection in one species, but severe or fatal disease in another species.

Avian chlamydiosis is a zoonotic disease that can affect people following exposure to air- or dustborne organisms when infected birds are in flocks or processed, or when organisms are shed from the digestive or respiratory tracts of infected birds confined in breeding aviaries, lofts, or wholesale or retail outlets.

Human disease most often results from exposure to psittacines or pigeons and can occur even if there is only brief proximity to a single infected bird. When workers are exposed to infected turkeys or ducks at processing, increased absenteeism due to acute respiratory disease often occurs ~1 wk after a flock with a high condemnation rate due to airsacculitis has been processed.

Some individuals, especially pregnant women and those with impaired immunity, are more susceptible than others. The illness in people is usually respiratory and characterised by abrupt onset of flu-like symptoms; pneumonia, organ failure, and death can result if the disease is severe or left untreated. Precautions should be taken when examining a dead infected bird (eg, detergent disinfectant to wet feathers, fan-exhausted examining hood, dust mask or plastic face shield, and gloves) to avoid exposure.

Etiology and Epidemiology

A recent taxonomic revision resulted in the causative organism being renamed Chlamydophila psittaci (formerly Chlamydia psittaci). The name of the disease resulting from infection with C psittaci remains avian chlamydiosis. C psittaci is an obligate intracellular bacterium. All strains of chlamydia share an identical genus-specific antigen in their lipopolysaccharide but often differ in the composition of other cell-wall antigens, providing a basis for serotypic identification. Currently, 8 serotypes are recognised; 6 (A-F) infect avian species and are distinct from mammalian chlamydia serotypes. Each avian serotype tends to be associated with certain types of birds. Serotype D is highly virulent for turkeys and can cause mortality of 30% or higher. Serotypes B and E are most frequently recovered from wild birds. Avian serotypes are capable of infecting people and other mammals.

Respiratory discharges or feces from infected birds contain elementary bodies that are resistant to drying and can remain infective for several months. Airborne particles and dust spread the organism. After inhalation or ingestion, elementary bodies attach to microvilli on mucosal epithelial cells and are internalised by endocytosis. Elementary bodies within endosomes in the cell cytoplasm differentiate into metabolically active, noninfectious reticulate bodies that divide and multiply, eventually forming numerous infectious, metabolically inactive elementary bodies. Newly formed elementary bodies are released from the host cell by lysis.

Possible sources of C psittaci include infected birds, asymptomatic carriers, vertical transmission from infected hens, infected rodents, and contaminated feed. Stressors and concurrent infections, especially those causing immunosuppression, can initiate shedding in latently infected birds and may cause recurrence of clinical disease. Carriers often shed the organism intermittently for extended periods. Persistence of C psittaci in the nasal glands of chronically infected birds may be an important source of organisms. Transmission is faecal-oral or by inhalation. The incubation period typically is 3-10 days but may be up to 2 mo in older birds or following low exposure.

Host and microbial factors, route and intensity of exposure, and treatment determine clinical course.

Clinical Findings and Lesions

Severity of clinical signs and lesions depends on the virulence of the organism and susceptibilty of the bird; asymptomatic infections are common. Nasal and ocular discharges, conjunctivitis, sinusitis, green to yellow-green droppings, fever, inactivity, ruffled feathers, weakness, inappetence, and weight loss can be seen in clinically affected birds.

Necropsy findings in acute infections include serofibrinous polyserositis (airsacculitis, pericarditis, perihepatitis, peritonitis), pneumonia, hepatomegaly, and splenomegaly. Multiple pale foci and/ or petechial hemorrhages can be seen in the liver and spleen. Similar lesions are seen in other systemic bacterial infections and are not specific for avian chlamydiosis. Multifocal necrosis in the liver and spleen is associated with large, granular, basophilic intracytoplasmic inclusions, occasional heterophils, and increased mononuclear cells (macrophages, lymphocytes, plasma cells) in hepatic sinusoids and splenic sinuses. Necrosis results from direct cell lysis or vascular damage. The latter is also the source of the generalised serofibrinous exudation. Enlargement and discoloration of the spleen or liver characterise chronic infections. Necrosis and inclusions are not seen, but the mononuclear cell response is present in these birds. Lesions are usually absent in latently infected birds, even though C psittaci is often being shed.

Diagnosis

Because of the variety of clinical presentations and common occurrence of latently infected carriers, no single diagnostic test can reliably determine infection. Procedures to detect the organism or antibodies are used. In general, the more acute the disease, the greater the number of infective organisms and the easier it is to make a diagnosis. When birds are acutely ill, clinical findings, including hematology, clinical chemistries, and radiology or typical gross lesions, are adequate for a tentative diagnosis. The organism can often be identified in impression smears of affected tissues stained by Giemsa, Gimenez, or Macchiavello's methods.

Antigen detection methods include immunohistochemistry (immunofluorescence, immunoperoxidase), ELISA, and PCR. Immunohistochemistry is accurate when done by a skilled person and the number of organisms is sufficient for detection. ELISA kits are available commercially and are relatively inexpensive, easy to use, and have good specificity, but low sensitivity.

They are most useful when birds are clinically ill. PCR tests have been developed but are not widely available and require further evaluation. Multiple samples collected for 3-5 days are recommended for detection of intermittent shedding by asymptomatic birds.

Confirmation requires isolation and identification of C psittaci in chick embryos or cell cultures (BGM, L929, Vero) at a qualified laboratory. Cloacal, choanal, oropharyngeal, conjunctival, or faecal swabs from live birds or tissues (eg, liver, spleen, serosal membranes) from dead birds should be submitted. Sampling from mutiple sites and over several days will increase detection of intermittent shedding. Freezing, drying, improper handling, and certain transport media can affect viability. Refrigeration; placing specimens in sealed plastic bags or other containers; using a special buffer prepared from sucrose, phosphate, and glutamase (SPG buffer); and prompt delivery of fresh specimens are preferred.

The laboratory should be contacted for directions on submitting samples before they are sent. Concurrent infections with other more easily diagnosed diseases (eg, colibacillosis, pasteurellosis, herpesvirus infections, mycotic diseases, etc.) may mask chlamydial infection. Laboratory and clinical findings should be correlated with each other. Chlamydiosis must be distinguished from other respiratory and systemic diseases of birds.

Antibodies may or may not be detectable depending on the test used, degree and stage of infection, and treatment of the bird. Interpretation of titres from single serum samples is difficult. A 4-fold increase in titres between paired acute and convalescent samples is diagnostic, and high titres in a majority of samples from several birds in a population are sufficient for a presumptive diagnosis. Serologic methods include direct and modified direct complement fixation, latex agglutination, elementary body agglutination, and direct and competitive ELISA. ELISA provides the greatest sensitivity and specificity compared with culture. The elementary body agglutination test detects IgM and is useful for determining recent infection.

Prevention and Treatment

Local governmental regulations should be followed wherever applicable. No effective vaccine for use in birds is available. Treatment will prevent mortality and shedding but cannot be relied on to eliminate latent infection; shedding may recur. Tetracyclines (chlortetracycline, oxytetracycline, doxycycline) are the antibiotics of choice. Drug resistance to tetracyclines is rare, but reduced sensitivity requiring higher dosages is becoming more common. Tetracyclines are bacteriostatic and only effective against actively multiplying organisms, making extended treatment times (from 2-6 wk, during which minimum-inhibitory concentrations in blood are consistently maintained) necessary.

Outbreaks in poultry flocks are not common. Treating infected flocks with chlortetracycline at 400-750 g/ton for a minimum of 2 wk before processing has effectively eliminated potential risk of infection for plant employees.

In companion birds, use of chlortetracycline-medicated feeds for 45 days is a standard recommendation for imported birds. Difficulties in palatability of the feed itself or high level of antibiotic necessary for adequate blood levels have limited its use. Long-acting oxytetracycline at 50-100 mg/kg, IM, every 2-3 days for 30 days, provides adequate continuous blood levels and results in elimination of shedding within 24 hr. However, muscle necrosis at injection sites may be extensive, which limits the usefulness of this treatment. Doxycycline in a formulation for IM use has been given at 75-100 mg/kg as a series of 7 injections over a 6 wk period. Addition of doxycycline to feeds can also result in adequate blood levels and has less effect on normal intestinal flora than does chlortetracycline. Supportive care for acutely affected birds also aids recovery.

Appropriate biosecurity practices are necessary for controlling the introduction and spread of chlamydiae in an avian population. Minimal standards include quarantine and examination of all new birds, traffic control to minimise cross-contamination, isolation and treatment of affected and contact birds, thorough cleaning and disinfection of premises and equiment (preferably with small units managed on an all-in/all-out basis), provision of uncontaminated feed, maintenance of records on all bird movements, and continual monitoring for presence of chlamydial infection.

The organism is susceptible to heat and most disinfectants (eg, 1:1,000 quaternary ammonium chloride, 1:100 bleach solution, 70% alcohol, etc.), but is resistant to acid and alkali. A voluntary cooperative improvement plan leading to certification of companion birds derived from chlamydia-free breeders has been developed.

Avian Nephritis Viral Infections

Avian nephritis viral infections are contagious infections of chickens and turkeys characterised by renal damage and visceral urate deposits, growth retardation, runting-stunting syndrome, and limited mortality (2-6%). They are seen mainly in chickens <7 days old, but interstitial nephritis can be observed in chicks up to 4 wk old. These infections have been reported worldwide. Subclinical infections are common and have been detected by serologic surveys in some SPF flocks.

Etiology and Transmission

The causal viruses are avian nephritis virus (ANV, an astrovirus), ANV-like viruses, and related enterovirus-like viruses (ELV). Strains vary in virulence and in antigenicity. Transmission occurs by direct or indirect contact. Indirect evidence suggests that egg transmission may occur. Infection can be transmitted by oral administration of virus to day-old birds. Virus is consistently isolated from the kidneys or the feces during the first 10 days after infection.

Nephropathogenic strains of infectious bronchitis virus (*Infectious Bronchitis: Introduction*) also cause interstitial nephritis. Therefore, when nephritis is diagnosed, it is necessary to isolate the causative agent.

Clinical Findings

Clinical signs vary from none to the so-called runting-stunting syndrome. Diarrhea and growth retardation are common in broilers.

Outbreaks with mortality of 2-6% can occur in chicks newly hatched up to 7 days old; cardinal necropsy findings are renal damage and visceral urate deposits (baby chick nephropathy).

Lesions

Nephritis is a common necropsy finding. Gross and microscopic lesions are often seen in the kidneys. Swelling, paleness, or yellowish discoloration with excessive urate deposition is frequent. Histologic lesions consist of a degeneration of the epithelial cells with infiltration of granulocytes, interstitial lymphocyte infiltration, and moderate fibrosis. In the latter stages, lymphoid follicles develop.

Some ELV induce only intestinal lesions varying from decreased length of the microvillus border to total desquamation of the intestinal epithelium.

Diagnosis

ANV and related viruses may be isolated by inoculation of suspected material (kidney or rectal contents) in the yolk sac of SPF chick embryos and in chick kidney cells. However, many ANV, ANV-like, and ELV viruses are difficult to isolate. The best method of detection is by electron microscopic examination of faecal preparations. Direct immunofluorescence performed on kidney sections is also a useful diagnostic procedure and allows quick differentiation from infectious bronchitis virus. Serologic diagnosis can be made using indirect immunofluorescence, seroneutralisation, or ELISA tests.

Treatment and Prevention

There is no effective treatment. General hygienic precautions are the only applicable preventive measures.

10

Avian Spirochetosis

Avian Borreliosis

Avian spirochetosis is an acute, febrile, septicemic, bacterial disease that affects a wide variety of birds.

Etiology, Epidemiology, and Transmission

The causal organism, Borrelia anserina, is an actively motile spirochete, ~0.2-0.3 μm × 8-20 μm, and consists of 5-8 loosely arranged coils. Cultivation in vitro is difficult. Borrelia will grow on Barbour-Stoenner-Kelly medium, but loses virulence after 12 passages. It can also be propagated in embryonating duck or chick embryos or in young ducks or chicks.

Spirochetosis is found in temperate or tropical regions, wherever the biologic vectors are found. The most common vector is Argas (Persicargas) persicus, the "cosmopolitan" fowl tick, but other Argas spp transmit the disease in different geographic areas. In the western USA, a highly efficient vector is A sanchezi.

Diverse immunologic and serologic types of B anserina have been demonstrated in many areas. Recovery from one type confers solid immunity against the homologous types for e"1 yr, but not against heterologous strains. Relapses, such as occur with some human Borrelia infections, are unknown in B anserina infection of birds; any reinfection can be attributed to a heterologous type.

Generally, an infected Argas tick can transmit the disease at every feeding and maintains the infection throughout larval, nymphal, and adult stages. The ticks also transmit the infection transovarially, ie, the F_1 larvae are infective. Ticks remain infected despite feeding on chicks hyperimmune to B anserina or on chicks with high blood levels of chemotherapeutic agents effective against Borrelia. Other vectors (lice, mosquitos, some species of ticks, inanimate objects) can transmit the spirochete mechanically to a susceptible host whenever the piercing apparatus becomes contaminated with blood that contains Borrelia. Ingestion of bile-stained faecal droppings containing the spirochete, contamination of feed or water, and cannibalism during spirochetemia can result in infection. After the bite of an infected tick, the incubation period is ~3-12 days.

Clinical Findings

Marked enlargement and mottling of the spleen is the most characteristic lesion. Signs are highly variable, depending on the virulence of the spirochete, and may be absent. Signs include listlessness, depression, somnolence, moderate to marked shivering, and increased thirst. Young birds are affected more severely than older ones. During the initial stages of the disease, there is usually a green or yellow diarrhea with increased urates. The course of the disease is 1-2 wk. Mild strains are common. However, in many tick-infested geographic areas, morbidity can approach 100% and mortality may be 33-77%. Egg production in layers or breeders may be reduced by 5-10%, with a higher number of small eggs.

Lesions

An enlarged spleen with petechial or ecchymotic hemorrhages, not unlike spleens in marble spleen disease of pheasants (*Hemorrhagic Enteritis of Turkeys and Marble Spleen Disease of Pheasants: Introduction*), is present. However, a contrasting situation may be seen in Mongolian pheasants, in which the spleen is reported to be small and pale. Occasionally, the liver may be swollen and contain focal areas of necrosis. Kidneys may be enlarged and pale. A green, catarrhal enteritis is common.

Diagnosis

Diagnosis depends on demonstration of Borrelia in the blood, either as actively motile during darkfield microscopy, or as stained spirochetes in Giemsa-stained blood smears. In young birds, the Borrelia may reach vast numbers per oil-immersion field and persist for several days. Older birds usually have low numbers of Borrelia that are detected only with difficulty, or not at all, and that persist for only 1-2 days. Anaemia is common and results in increased numbers of immature RBC. Agar-gel diffusion and various serologic tests have been described but are of questionable value due to diverse serotypes that exist in some localities. Specific agglutinins clump the spirochetes in successively larger clumps during the terminal stages of the disease. Agglutination-lysis then begins to disintegrate these clumps, and spirochetal degradation products are liberated, which may result in pyrexia. Death occurs most often 1-3 days after Borrelia disappear from the bloodstream. Spirochetal antibodies are readily detected in yolks of eggs laid by infected hens.

Treatment and Control

Several antibacterial agents are effective. The most widely used are penicillin derivatives, but the streptomycins, tetracyclines, and tylosin are also effective. The antibiotics can be completely efficacious if begun when the number of spirochetes per oil-immersion field is low or moderate; however, if large numbers of spirochetes are present in the bloodstream, the sudden liberation of large quantities of spirochetal degradation products can result in higher mortality than no treatment. Control must be directed against the biologic vector. Argas ticks are notable for their long lifespan, ability to survive for extended periods without a blood meal, efficiency in transmitting the spirochete, and ability to remain securely hidden in cracks and crevices often beyond the effective reach of pesticides. Accordingly, control is difficult. A combination of tick eradication and immunisation is the most effective means of control.

Immunisation can be highly successful and, next to eradication of the biologic vector, is the preferred method of control. Bacterins prepared from local strains of Borrelia have been used with success. Vaccines may be prepared from formalin- or phenol-inactivated material from lysates of blood, tissues, embryos, or eggs infected with B anserina, and may be lyophilised or liquid. Whole-egg propagated bacterins are usually given in 1 or 2 IM injections. Little if any cross-protection is afforded to different serotypes. Birds normally have protective immunity after recovering from natural infection.

Listeriosis

Listeriosis is quite rare in birds and usually occurs as a septicemia or sometimes as a localised encephalitis. Encephalitis combined with septicemia has been seen in young geese. Chickens, turkeys, geese, ducks, canaries, and parrots appear to be the most commonly affected avian species. In workers at poultry-processing plants, conjunctivitis due to Listeria monocytogenes has been linked to handling of apparently normal but infected chickens.

Human infections have also resulted from consumption of contaminated poultry or poultry products. Abortions and congenitally infected babies have been associated with handling of L monocytogenes -positive birds or those that have died with the disease, but these cases were not confirmed.

Etiology and Epidemiology

L monocytogenes is a gram-positive, coccoid to bacillus-shaped, nonsporeforming bacteria that tends to form long filaments, particularly in older cultures. Based on somatic and flagellar antigens, several serotypes have been described. L monocytogenes can be cultured on blood and tryptose agar or brain-heart infusion. It is widely distributed among avian species.

The organism is common in feces and soil, with numbers increasing in late winter and early spring. It has been isolated from apparently normal birds and from birds dying of causes other than uncomplicated listeriosis; therefore, it is possible that carrier birds play an important role in the perpetuation of the disease in birds and mammals. It is commonly associated with concurrent diseases such as coccidiosis, infectious coryza, salmonellosis, and parasitic infections demonstrating the largely opportunistic character of the organism.

Clinical Findings

Young birds appear to be more susceptible than mature ones. Transmission and subsequent infections occur by ingestion of contaminated nasal secretions, feces, and soil. Infection can also occur via inhalation and wound contamination.

In most avian species, the incubation period has not been documented; in turkeys, it is 16 hr to 52 days. Frequently, L monocytogenes infections are subclinical. Chickens and turkeys are relatively resistant to natural infection. However, signs of infection are suggestive of a septicemia and include depression, listlessness, and peracute death. In this form, it is common to find only dead birds. In the subacute and chronic forms, signs are related to encephalitis and include torticollis, stupor, paresis, and paralysis. Adult birds may die suddenly with septicemia, while young birds tend to have chronic infections. Emaciation and diarrhea are seen in some affected birds.

Lesions

In uncomplicated listeriosis, lesions include multiple areas of degeneration and necrosis of the myocardium with congestion, increased pericardial fluid, and pericarditis. Petechial hemorrhages can be seen in the proventriculus and heart. Splenomegaly and hepatomegaly with bile retention and focal areas of necrosis are common. In the encephalitic form, no gross brain lesions are seen; microscopically, however, gliosis in the cerebellum with microabscesses containing gram-positive bacteria are present in the midbrain and medulla.

Diagnosis

Listeriosis can be suspected based on the history, clinical signs, necropsy lesions, and microscopic observation of the bacteria in the myocardial fibrils, hepatocytes, or both. The diagnosis can be confirmed by isolation from the blood, liver, heart, spleen, or brain of a gram-positive, nonacid-fast, nonsporeforming bacillus that is catalase-positive, motile, aerobic, and that ferments sugars.

Isolation by direct culture of the affected tissues may not be successful because of low concentration of organisms in the tissues; however, recovery increases significantly if a portion of the specimen is refrigerated for 4-8 wk and subcultured weekly. Chick embryos are readily infected and can be used for organism identification. Differential diagnoses include colibacillosis, pasteurellosis, erysipelas, velogenic viscerotropic Newcastle disease, and many other acute and chronic bacterial diseases.

Treatment and Control

The organism is often resistant to many of the commonly used antibiotics. However, the tetracyclines have been efficacious in both the acute and subacute forms when given at 25 mg/kg, PO, sid for 1 wk. Treatment of the chronic form is usually unsuccessful. Widespread use of antimicrobials in the feed for growth promotion may have prophylactic value. Rigid sanitation and disinfection procedures with culling and isolation of affected birds may be helpful. Prevention should focus on identifying and eliminating the source of infection.

Malabsorption Syndrome

Runting-stunting Syndrome, Pale Bird Syndrome

Malabsorption syndrome is characterised by stunted growth and a lack of skin pigmentation in growing chickens, most commonly meat type or broilers. It has been identified in virtually all countries in which intensive poultry production occurs.

Etiology and Transmission

Mycotoxins and several viruses, including enteroviruses, parvoviruses, astroviruses, caliciviruses, arenaviruses, and reoviruses have been implicated. Although the etiology is believed to be complex, only mycotoxins, enteroviruses, and reoviruses have thus far been identified as potential etiologic factors.

The specific feedborne mycotoxins involved and concentration needed to induce the syndrome are not well understood. Numerous enteric viruses are prevalent worldwide in commercially produced poultry. Transmission of viruses occurs via faecal-oral routes.

Clinical Findings

The disease is typically recognised in broiler chicks 1-3 wk old. It is characterised by stunted growth; lack of pigmentation in the skin, feet, or beak; slow feathering; broken or twisted feathers; undigested feed in the feces; and/or poor feed conversion ratios. Diarrhea is common during the initial phases. Severely affected birds do not respond immediately to changes in feed or management practices and are usually culled from flocks before processing.

Lesions

The severity and type of lesions resulting from both field and laboratory infections vary with the particular agents or combinations of agents involved. Lesions often include enlarged proventriculi, small gizzards, and orange mucus in the small-intestinal lumen. No consistent microscopic lesions are found. Encephalomalacia or rickets may be seen occasionally, presumably as a result of malabsorption or malassimilation of nutrients.

Diagnosis

Clinical signs and lesions permit a presumptive diagnosis, although a similar gross appearance can be caused by a retrovirus. More conclusive diagnostic evidence includes finding either viruses in the lesions or dietary toxins.

Prevention and Control

There is no effective treatment for severely affected birds. Sanitation and disinfection will reduce the burden of challenge caused by multiple infectious organisms. There are no vaccines that will prevent malabsorption syndrome. Reovirus vaccines can prevent the stunting and poor feed conversions that occur with pathogenic reovirus infections. Feeds should be analysed for dietary toxins, and high levels of toxins should not knowingly be fed to commercial poultry.

Mycoplasmosis

Several Mycoplasma spp have been isolated from avian hosts; M gallisepticum, M iowae, M meleagridis, and M synoviae are the most important. Mycoplasmas are fastidious bacteria, 0.3-0.8 μm in diametre; they lack a cell wall and require a rich growth medium containing serum. They do not survive for more than a few days outside the host and are vulnerable to common disinfectants. Each has distinctive epidemiologic and pathologic characteristics.

Mycoplasma Gallisepticum Infection

M gallisepticum infection is commonly designated as chronic respiratory disease in chickens and as infectious sinusitis in turkeys. Infection may also be seen in pheasants, chukar partridges, and peafowl. Infection in pigeons, quail, ducks, geese, and psittacine birds should be considered. Passerine-type birds are quite resistant, although M gallisepticum is the major cause of natural outbreaks of conjunctivitis in wild house finches (Carpodacus mexicanus) in the eastern USA. The disease is worldwide. Its effects are most severe in large commercial operations during winter.

M gallisepticum is the most pathogenic avian mycoplasma; however, strains may differ markedly in virulence. Primary isolation is made in enriched broth medium containing 10-15% serum, then plated on agar. Typical colonies are identified by immunofluorescence.

Transmission, Epidemiology, and Pathogenesis

In the USA, most breeder flocks are free of M gallisepticum, and outbreaks are due to lateral transmission from infected chickens; however, in some parts of the world, egg transmission is a major source of infection. The incidence of egg transmission is highly variable, ranging up to 30-40% during the first 2 mo after infection of susceptible birds in production. The transmission rate then lessens and is inconsistent (0-5%) until the end of production. Birds infected before the onset of production transmit through the egg at a much lower rate, if at all. The infection may be dormant in the infected chick for days to months, but when the flock is stressed, aerosol transmission occurs rapidly and infection spreads through the flock. Live virus vaccination, natural virus infection, cold weather, or crowding may initiate the spread. In addition, the infection may be carried by personnel (especially from an infected to a clean flock), fomites, or introduction of infected birds. In many flocks, the source of infection cannot be determined.

The epithelium of the upper air passages is most susceptible to infection; however, in severe, acute disease the infection is also found in the lower respiratory tract. There is a marked interaction between respiratory viruses, Escherichia coli, and M gallisepticum in the pathogenesis of chronic respiratory disease. Once infected, birds remain carriers for life.

Clinical Findings

In chickens, infection may be inapparent or result in varying degrees of respiratory distress, with slight to marked rales, difficulty breathing, coughing, and/or sneezing. Morbidity is high and mortality low in uncomplicated cases. Nasal discharge and frothiness about the eyes may be present. In turkeys, the disease is generally more severe than in chickens, and swelling of the paranasal sinus is common. Feed efficiency and weight gains are reduced. Broilers and market turkeys may suffer high condemnations at processing due to airsacculitis. In laying flocks, birds may fail to reach peak egg production, and the overall production rate is lower than normal.

Lesions

Uncomplicated M gallisepticum infections in chickens result in relatively mild sinusitis, tracheitis, and airsacculitis. E coli infections are often concurrent and result in severe air sac thickening and turbidity, with exudative accumulations, fibrinopurulent pericarditis, and perihepatitis, particularly in broilers. Turkeys develop severe mucopurulent sinusitis and varying degrees of tracheitis and airsacculitis. The mucous membranes are thickened, hyperplastic, necrotic, and infiltrated with inflammatory cells. Lymphofollicular areas are found in the submucosa.

Diagnosis

Agglutination reactions and ELISA are commonly used for diagnosis. M gallisepticum should be confirmed by isolation and identification, PCR, or hemagglutination-inhibition because nonspecific false agglutination reactions are common, especially after the inoculation of inactivated, oil-emulsion vaccines or M synoviae infection. Isolates must be identified, because birds may also be infected with nonpathogenic Mycoplasma spp. PCR is commonly used to rapidly detect the organism in upper respiratory tissues. Newcastle disease, infectious bronchitis, influenza, and other respiratory pathogens should be considered in the differential diagnosis.

Treatment and Control

In the field, many cases of M gallisepticum infection are complicated by other pathogenic bacteria; thus, effective treatment must also attack the secondary invader. Most strains of M gallisepticum are sensitive to a number of antibiotics, such as chlortetracycline, erythromycin, oxytetracycline, spectinomycin, tiamulin, tylosin, or a fluoroquinolone such as enrofloxacin. Antibiotic is usually given in the feed or water for 5-7 days; however, in turkeys, antibiotic may be given initially by injection, followed by feed or water medication. Antibiotics may alleviate the clinical signs and lesions but do not eliminate infection. Eradication of M gallisepticum from chicken and turkey breeding stock is well advanced in the USA and several other countries. The most effective control program is to identify breeders without serum agglutination or ELISA titres and to maintain seronegative stock. In valuable breeding stock, treatment of eggs, usually with tylosin or heat, may be used to eliminate egg transmission to progeny. Medication is not a good long- term control method but is of value in treating individual infected flocks.

The use of birds free of M gallisepticum is desirable, but infection in multiple-age commercial egg farms where depopulation is not feasible is a problem. An inactivated, oil-emulsion bacterin is available in most countries; it prevents egg production losses but not infection. A live vaccine has been licensed in the USA for use in infected, multiple-age layer flocks but may be used only with permission of the state veterinarian. The vaccine consists of a mild strain of M gallisepticum (F-strain) and is usually given at ~10-14 wk of age. F-strain is of low pathogenicity for chickens but is fully virulent for turkeys. Vaccinated birds remain carriers, and immunity lasts through the laying season. Recently, 2 nonpathogenic live vaccine strains (6/85 and ts-11) have been introduced; these strains offer the advantage of improved safety and are in widespread use in commercial layers.

Mycoplasma Iowae Infection

M iowae was originally thought to be of low pathogenicity in producing air sac lesions in chickens and turkeys, but it is a potentially important cause of reduced hatchability in turkeys. Antigenicity and pathogenicity vary considerably among M iowae strains. M iowae is resistant to 1% bile salts, and an enriched medium similar to those used for other avian mycoplasmas is suitable.

Infection was common in turkey flocks in Europe and North America, but the infection rate has now been reduced by intensive eradication efforts in breeding stocks. It is a relatively uncommon infection of chickens. M iowae is egg transmitted, but little is known of other aspects of its epidemiology.

Many strains of M iowae are lethal to turkey embryos. After experimental inoculation of young poults, stunting, poor feathering, and various skeletal deformities such as tenosynovitis and chondrodystrophy develop, but the mechanism is unknown. These effects have not been recognised in the field, probably because most infected birds die before hatching. Older birds appear to be quite resistant.

Clinical Findings, Lesions, and Diagnosis

Affected turkey breeder flocks show no clinical signs other than reduced hatchability (usually 2-5%). In many flocks, the hatchability returns to normal after 1-2 mo.

Most embryos die during the mid to late stages of incubation. Dead turkey embryos are edematous, congested, and stunted; they may have clubbed down. Poults challenged in ovo or at 1 day of age may develop various skeletal deformities such as rotated tibia, deviated toes, chondrodystrophy, or erosion of the articular cartilage of the hock joint. Feathers may also be poorly developed. Chicks challenged at 1 day of age may develop tenosynovitis and ruptured tendons.

Turkeys apparently have a poor antibody response, and no reliable serologic test is available. Diagnosis relies on isolation and identification of the causative agent.

Treatment and Control

The best method of control is to maintain flocks free of M iowae; however, because serology is unreliable, this may be difficult. Dipping hatching eggs in solutions of enrofloxacin has significantly reduced losses in hatchability.

Mycoplasma Meleagridis Infection

M meleagridis infection is a widespread, egg-transmitted disease of turkeys found worldwide. The primary lesion in the progeny is airsacculitis. M meleagridis is thought to be a specific pathogen for turkeys, and the organism is commonly found in the respiratory and reproductive tracts. It has been eradicated in most basic breeder and many commercial flocks.

M meleagridis was recognised as a pathogen of turkeys after widespread elimination of the bacteria from breeding stock.

Transmission and Pathogenesis

Infection is established primarily through egg transmission, which can be as high as 30-50% or higher early in the production cycle. However, transmission of M meleagridis is also related to genital contact. Early infections usually become quiescent at sexual maturity. In the tom, the phallus and adjacent tissues are infected and contaminate semen, thus infecting the vagina of the hen. Hens may retain infection in the bursa of Fabricius, which serves as a source of infection of the reproductive tract after rupture of the cloacal-vaginal occluding membrane at puberty.

Infection ascends the reproductive system and may reach the surface of the ovary. The high rate of egg transmission of M meleagridis is from infection of the reproductive tract being incorporated into the egg after ovulation. Infection of the respiratory tract leads to transmission among birds in young flocks and may be a factor in the spread to flocks previously free of infection. Hatchery transmission is also possible.

The marked difference in the pathogenicity of various strains of M meleagridis results in variable clinical manifestations. The high incidence of air sac infection in poults suggests a symbiotic host-parasite relationship. M meleagridis may be involved in crooked necks and leg deformities, but the pathogenesis of this syndrome is not clear. The vaginas of naturally infected hens are free of infection 1-3 mo after the source of infection is removed. Immunity is not permanent, and hens can be reinfected with contaminated semen.

Clinical Findings

Embryo infection appears to reduce hatchability, poult quality, and growth rate. Superimposed stress may cause considerable mortality in poults during the first few weeks. Infection during early rapid growth of hock joints, periarticular tissues, cervical vertebrae, and adjacent bone may produce major bone deformities such as crooked necks and hocks. Rales may develop in poults 3-8 wk old and persist for several weeks without significant mortality or serious interference with growth.

Lesions

Day-old poults have thoracic airsacculitis with thickening, turbidity, and marked caseous exudate. In 1-3 wk, the lesions may extend to the abdominal air sacs. These lesions recede with age. The air sac lesions of roaster and mature birds are probably related to other factors. Tracheitis may be present, but sinusitis does not occur.

Microscopic lesions in hens consist of lymphocytic foci in the fimbria, uterus, and vagina. In young poults, inflammatory lesions are seen in the air sacs and lungs.

Diagnosis

A high incidence of air sac lesions in day-old poults suggests M meleagridis infection. The serum plate agglutination or ELISA test may be used. Confirmation is generally by hemagglutination-inhibition, isolation and identification of the organism, or both. M gallisepticum, chlamydiae, bacteria, and respiratory viruses such as influenza must be eliminated as causes of air sac infection.

Treatment and Control

The commercial use of flocks free of M meleagridis should be monitored by serology and/or by examining pipped embryos or cull poults for airsacculitis. Semen used for insemination must be free of M meleagridis.

Dipping eggs in tylosin or another suitable antibiotic reduces the incidence of transmission in infected flocks and may improve weight gains and feed conversion ratios. Inoculation SC of a suitable antibiotic at 1 day of age or water medication for the first 5-10 days appears to reduce airsacculitis caused by M meleagridis and may improve weight gain.

Mycoplasma Synoviae Infection

M synoviae was first recognised as an acute to chronic infection of chickens and turkeys that produced an exudative tendinitis and bursitis; it now occurs most frequently as a subclinical infection of the upper respiratory tract. M synoviae infection is also a complication of airsacculitis in association with Newcastle disease or infectious bronchitis. It is seen primarily in chickens and turkeys, but ducks, geese, guinea fowl, parrots, pheasants, and quail may also be susceptible. Serum (preferably swine serum) and nicotinamide adenine dinucleotide are required for growth on artificial media.

Transmission, Epidemiology, and Pathogenesis

M synoviae is egg-transmitted, but the rate is low (probably <5%), and some hatches of progeny may be free of infection. Egg transmission is greatest during the first 1-2 mo after infection of susceptible breeders. Lateral transmission is similar to that of M gallisepticum, but the rate of spread is generally more rapid.

M synoviae isolates vary widely in pathogenicity. Isolates from cases of airsacculitis are more apt to produce air sac lesions than isolates from synovial fluid or membranes. Some strains produce the typical clinical disease of synovitis. The paucity of natural outbreaks of clinical synovitis in chickens in recent years may be related to the adaptation of M synoviae to the respiratory tract; however, clinical synovitis in turkeys is relatively common.

Clinical Findings

Although slight rales may be present in birds with respiratory infection, usually no signs are noticed. Younger birds, especially those under stress or suffering concurrent infections, are more likely to be affected. Outbreaks of infectious synovitis occur most commonly in chickens at 4-6 wk and in turkeys at 10-12 wk. Lame birds tend to sit. The more severely affected birds are depressed and are found around the feeders and waterers. Swellings of the hocks and footpads are seen. Morbidity is 2-15%, and mortality 1-10%. The effect on egg production is minimal, but instances of egg production losses have occurred.

Lesions

In the respiratory syndrome, airsacculitis occurs when the bird is stressed from Newcastle disease, infectious bronchitis, or improper ventilation. In many cases, air sac lesions resolve after 1-2 wk. Early in synovitis, the liver is enlarged and sometimes green. The spleen is enlarged, and the kidneys are enlarged and pale. A yellow to gray, viscid exudate is present in almost all synovial structures; it is most commonly seen in the keel bursa, hock, and wing joints. In chronic cases, this exudate may become inspissated and orange.

Diagnosis

A presumptive diagnosis can be based on the lesions and clinical signs, but laboratory confirmation is necessary. Skeletal abnormalities must be eliminated as the cause of lameness. The disease must be differentiated from viral tenosynovitis and from staphylococcal and other bacterial infections.

The serum plate agglutination or ELISA test is used to detect infected flocks, but cross-reactions with M gallisepticum and other nonspecific reactions may occur. Reactors are confirmed as positive by hemagglutination-inhibition or by isolation and identification of the organism. PCR may be used to rapidly detect the organism in infected tissues. In turkeys, the agglutination test for M synoviae may not be reliable.

Treatment and Control

Serologic testing and isolation similar to those for M gallisepticum have resulted in eradication of the infection in most primary breeder flocks of chickens and turkeys. Administration of a tetracycline antibiotic in the feed may be beneficial in treatment or prevention of synovitis. When airsacculitis is a problem, preventive antibiotic therapy during the time of respiratory reaction to Newcastle disease and infectious bronchitis vaccine may be helpful. Medication of breeder flocks is of little value in preventing egg transmission.

Salmonelloses

Historically, the 3 salmonellae infections in poultry causing severe economic losses are Salmonella pullorum, S gallinarum, and S arizonae. Through the institution of control programs, the incidence of infections with these salmonellae has decreased dramatically. In addition to the above salmonellae, S paratyphoid infections in poultry are relatively common and have public health significance because of the consumption of contaminated poultry products. S pullorum and S gallinarum are highly host-adapted to chickens and turkeys. S arizonae is most important in turkeys, with chickens occasionally affected. There are ~2,000 nonhost-adapted species (paratyphoid) that may be transmitted to almost all animals.

Pullorum Disease

Etiology and Transmission

Infections with Salmonella pullorum usually cause very high mortality (potentially approaching 100%) in young chickens and turkeys. In adult chickens mortality may be high, but frequently there are no clinical signs. Pullorum disease was once common but has been eradicated from most commercial chicken stock, although it may occur in other avian species (eg, guinea fowl, quail, pheasants, sparrows, parrots, canaries, and bullfinches). Infection in mammals is rare.

Transmission is primarily through the egg but also occurs via direct or indirect contact with infected birds. Infection transmitted via egg or hatchery contamination usually results in death during the first few days of life up to 2-3 wk of age.

Clinical Findings and Lesions

Affected birds huddle near a heat source, are anorectic, appear weak, and have whitish faecal pasting around the vent (diarrhea). Survivors frequently become asymptomatic carriers with localised infection of the ovary. Some of the eggs laid by such hens hatch and produce infected progeny. Lesions in young birds usually include unabsorbed yolk sacs and classic gray nodules in the liver, spleen, lungs, heart, gizzard, and intestine. Firm, cheesy material in the ceca (cecal cores) and raised plaques in the mucosa of the lower intestine are sometimes seen. Occasionally, synovitis is prominent. Adult carriers usually have no gross lesions but may have nodular pericarditis, fibrinous peritonitis, or hemorrhagic, atrophic regressing ovarian follicles with caseous contents. In mature chickens, chronic infections produce lesions that are indistinguishable from those of fowl typhoid.

Diagnosis

Lesions may be highly suggestive, but diagnosis should be confirmed by isolation, identification, and serotyping of S pullorum. Infections in mature birds can be identified by serologic tests, followed by necropsy evaluation complemented by microbiologic culture and typing for confirmation. Official testing recommendations are outlined in the USA National Poultry Improvement Program (NPIP).

Treatment and Control

Treatment of infected flocks to alleviate the perpetuation of the carrier state is not recommended. Control is based on routine serologic testing of breeding stock to assure freedom from infection.

Fowl Typhoid

Etiology and Epidemiology

The causal agent is Salmonella gallinarum. The incidence of fowl typhoid is low in the USA and Canada, but much higher in other countries. Although S gallinarum is egg-transmitted and produces lesions in chicks and poults similar to those produced by S pullorum, there is a much greater tendency to spread among growing or mature flocks. Mortality in young birds is similar to S pullorum infection but may be higher in older birds.

Clinical Findings and Lesions

Clinical signs and lesions in young birds are similar to those of infection with S pullorum. The older bird may be pale, dehydrated, and have diarrhea. Lesions in the older bird may include a swollen, friable, and often bile-stained liver, with or without necrotic foci, enlarged spleen and kidneys, anaemia, and enteritis.

Diagnosis

Diagnosis should be confirmed by isolation and identification and serotyping of S gallinarum (NPIP testing procedure).

Treatment and Control

Treatment and control are as for pullorum disease. There are no federally licensed vaccines in the USA. In other countries, vaccines (killed or modified live) made from a rough strain of S gallinarum (9R) have been useful in controlling mortality. More recently, vaccines derived from outer membrane proteins, mutant strains, and a virulence-plasmid-cured derivative of S gallinarum have shown promise in protecting birds against challenge. The standard serologic tests for pullorum disease are equally effective in detecting fowl typhoid.

Arizona Infection

Etiology, Clinical Findings, and Lesions

Many serotypes have been identified from various birds, mammals, and reptiles. Foodborne infections occasionally occur in humans. Reptiles, wild birds, rats, and mice are frequently infected and are thought to act as a reservoir of infection for poultry. Clinical signs and lesions are not distinctive. Mortality is usually limited to the first 3-4 wk of age, and adult carriers may not develop appreciable clinical signs. Infection tends to persist in a flock. Poults are diarrheic, listless, and unthrifty; in some flocks, birds may develop eye opacity and blindness. Neurologic signs occur due to infection of the brain.

Lesions include unabsorbed yolk sacs, enlarged and mottled livers, congestion in the duodenum, and caseous cecal cores. Some birds develop peritonitis, salpingitis, local ovarian infections, or ophthalmitis. Purulent material may be seen in the meninges.

Diagnosis

Diagnosis is based on isolation and identification of the organism. Affected eyes and brain are excellent sites for isolation. Environmental samples also may be used for detecting the microbe. Egg transmission levels are often high; therefore, culture of dead embryos, eggshells, and cull poults may identify infected breeding stock. Effective serologic tests have not been developed.

Treatment and Control

Killed vaccines (bacterins) have been used in infected breeder flocks to reduce egg transmission and to develop breeding flocks free of S arizonae. Early fumigation of hatching eggs and rigorous hatchery sanitation also aid in reducing transmission. Antibiotics are given by injection to day-old poults to minimise mortality. Birds may still carry and shed the organism even after treatment.

Paratyphoid Infections

Etiology, Clinical Findings, and Lesions

Paratyphoid infections can be caused by any one of the many nonhost-adapted salmonellae. These *Salmonella* may infect many types of birds, mammals, reptiles, and insects. Paratyphoid infections are of public health significance via contamination and mishandling of poultry products. Salmonella typhimurium, S enteritidis, and S heidelberg are among the most common salmonella infections in poultry, although infections may be produced by 10-20 different serotypes in the USA. Some species or strains are more pathogenic than others. The prevalence of other species varies widely by geographic location and season. Transmission usually occurs horizontally from infected birds, contaminated environments, or infected rodents. With the exception of S enteritidis, transmission of most serotypes to progeny from infected breeders is mainly through faecal contamination of the eggshell. Infected birds remain carriers.

Clinical Findings and Lesions

Although not common, clinical signs are sometimes seen in young birds. Mortality is most often limited to the first few weeks of age. Depression, poor growth, weakness, diarrhea, and dehydration are hallmarks of the disease, although these clinical signs are not distinctive. Lesions may include an enlarged liver with focal necrosis, unabsorbed yolk sac, enteritis with necrotic lesions in the mucosa, and cecal cores. Infections occasionally localise in the eye or synovial tissues. Conversely, there may be no lesions due to acute death caused by septicemia. Isolation, identification, and serotyping of the causal agent are essential for diagnosis. Serology is not highly reliable.

Treatment and Control

General control measures for the paratyphoid Salmonella include strict sanitation in the hatchery, fumigation of hatching eggs, pelleting of feed to destroy salmonellae, cleaning and disinfection of poultry houses, rodent control, and use of competitive exclusion products. Several antibacterial agents help prevent mortality but cannot eliminate flock infection. Maintenance of poultry in confinement and exclusion of all pets, wild birds, and rodents help prevent introduction of infection.

Salmonella enteritidis (a paratyphoid Salmonella serotype) is a major food safety concern, primarily for the egg-laying industry. Possible sources in commercial layers include transmission from breeders, contaminated environments, infected rodents, and contaminated feed. Transmission to progeny from breeders is mainly through eggshell contamination, although, unlike other paratyphoid Salmonella, transovarial transmission may also occur. The NPIP now includes S enteritidis control measures in breeders, including depopulation of infected flocks, cleaning and disinfection of pullet and layer houses, extensive and improved rodent control programs, use of competitive exclusion products, vaccination, and proper handling and refrigeration of eggs.

Tuberculosis

Tuberculosis is a slowly spreading, chronic, granulomatous bacterial infection, characterised by gradual weight loss. All birds appear to be susceptible, although to variable degrees; pheasants seem to be highly susceptible, while the disease is uncommon in turkeys. Tuberculosis is more prevalent in captive than in wild birds. Tuberculosis is unlikely to occur in commercial poultry due to the short life span and husbandry practices used.

Etiology and Epidemiology

Mycobacterium avium *var avium* is the cause. Serologic identification of isolates is essential to differentiate strains of M avium that cause disease in chickens and birds (serovars 1, 2, and 3) from other serovars that fail to produce disease in these species. M tuberculosis has infrequently been isolated from parrots and canaries. M avium is very resistant; it can survive in soil for up to 4 yr, in 3% hydrochloric acid for e"2 hr, and in 4% sodium hydroxide for e"30 min. Tuberculosis is found worldwide, most commonly in small, barnyard flocks and in zoo aviaries; it is rarely found in young flocks. Wild birds, such as cranes, sparrows, starlings, and raptors, have been found to be infected. Tuberculosis has been found in emus and other ratites.

Infected birds with advanced lesions excrete the organism in their feces. Cadavers and offal may infect predators and cannibalistic flockmates. Rabbits, pigs, and mink are readily infected.

Cattle exposed to contaminated feces may respond to mammalian tuberculin and to johnin. M avium may cause disease in humans; serovar 1, often isolated from tuberculous chickens, has been isolated from people with acquired immunodeficiency syndrome.

Clinical Findings and Diagnosis

Signs usually do not develop until late in the infection when birds become thin and sluggish, and lameness may be seen. In chickens, granulomatous nodules of varying size are usually found in the liver, spleen, bone marrow, and intestine. Some exotic species may have lesions in the liver and spleen without intestinal involvement, but bone marrow and small mesenteric nodules may be found. Lesions are not calcified.

Live birds may be tested with avian tuberculins, although these are of little value in birds that do not have wattles. Large numbers of acid-fast bacteria in smears from lesions provide a tentative diagnosis.

Control

Chemotherapy is ineffective. In commercial poultry flocks, relatively rapid turnover of populations, together with improved general sanitation, has largely eliminated this once common infection. Infected poultry should be destroyed, and housing facilities thoroughly cleaned and disinfected using cresylic compounds.

Dirt-floored houses should have several inches of the floor removed and replaced with dirt from a place where poultry have not been maintained. All openings should be screened against wild birds. Avian tuberculosis in zoos is difficult to eradicate. New additions to the aviary should be quarantined for 2-3 mo. The movement of ratites through sales and the long life of these animals have made tuberculosis a major concern for ratite producers. Isolation of ratites purchased at sales is essential to prevent the introduction of tuberculosis into established flocks.

Miscellaneous Conditions of Poultry

Ascites Syndrome

Ascites is an accumulation of noninflammatory transudate in one or more of the peritoneal cavities or potential spaces. The fluid, which accumulates most frequently in the 2 ventral hepatic, peritoneal, or pericardial spaces, may contain yellow protein clots. Ascites may result from increased vascular hydraulic pressure, vascular damage, increased tissue oncotic pressure, decreased vascular oncotic (usually colloidal) pressure, or blockage of lymph drainage.

The most common cause of ascites is increased vascular hydraulic pressure in the venous system, which is most commonly caused by right ventricular failure (RVF) or hepatic fibrosis.

In poultry, RVF is usually secondary to valvular insufficiency and may result from inflammatory (myocarditis, valvular endocarditis) or degenerative disease of the myocardium or valves or from congenital heart disease. In turkeys, spontaneous cardiomyopathy (*Round Heart Disease of Turkeys: Introduction*) is a common cause of ascites. However, the most common cause of ascites in meat-type chickens is RVF in response to increased pulmonary arterial resistance. Pulmonary hypertension occurs frequently in chickens secondary to the hypoxia of altitude with resultant polycythemia and increased blood viscosity. It also occurs frequently secondary to the red blood cell rigidity of sodium toxicity and less frequently from lung pathology. When ascites occurs at low altitudes in meat-type chickens, which have a high metabolic oxygen requirement, it is usually caused by primary or spontaneous pulmonary hypertension because of insufficient capacity of the pulmonary capillaries.

In poultry, liver damage may be caused by aflatoxin or by toxins from plants such as Crotalaria. In broiler chickens, obstructive cholangiohepatitis (caused by Clostridium perfringens infection) is the most common cause of the liver damage, which results in ascites. In both meat-type ducks and breeders, amyloidosis of the liver frequently causes ascites.

Pathogenesis and Epidemiology

Pulmonary hypertension syndrome (PHS) is caused by increased pressure in the pulmonary arteries when the heart tries to pump more blood through the lungs to meet the body's oxygen requirement. The resultant volume and pressure overload on the right ventricle cause dilatation and hypertrophy of the right ventricular wall, valvular insufficiency, RVF, and ascites. Bird lungs are rigid and fixed in the thoracic cavity. The capillaries can expand very little to accommodate increased blood flow.

Lung size in proportion to body weight, and particularly to muscle mass, decreases as meat-type chickens grow. Increased blood flow results in primary pulmonary hypertension and cor pulmonale with sporadic cases of RVF and ascites in fast-growing broilers.

Predisposing factors that increase oxygen demand (e.g., cold), reduce oxygen-carrying capacity of the blood (e.g., acidosis, carbon monoxide), increase blood volume (e.g., sodium), or interfere with blood flow through the lung (e.g., lung pathology that narrows or occludes capillaries, increased RBC rigidity, or polycythemia with increased blood viscosity) may result in flock outbreaks of PHS with or without ascites.

The incidence of PHS is >2% in some broiler and many roaster flocks and is occasionally 15-20% in other roaster flocks. Right ventricular hypertrophy is the response to an increased workload and eventually leads to RVF if the volume or pressure load persists. Hypertrophy of the right ventricular wall is directly related to pulmonary hypertension, and the ratio of the right ventricle to the total ventricular mass can be used as a measure of the increased pressure load on the right ventricle.

Clinical Findings

Occasionally, young broilers develop PHS, particularly if increased sodium or lung pathology (e.g., aspergillosis) is involved, but in primary pulmonary hypertension, mortality is greatest after 5 wk of age. Clinical signs are not seen until RVF occurs and ascites develops. Clinically affected broilers are cyanotic, the abdominal skin may be red, and peripheral vessels congested.

Because growth stops as RVF develops, affected broilers may be smaller than their pen mates. However, rapid growth rate is a known predisposing factor, and sometimes the largest broilers are affected, with occurrence in males more frequent than in females. The ascites increases the respiratory rate and reduces exercise tolerance. Affected broilers frequently die on their backs, and differential diagnosis includes flip-over disease (*Flip-over Disease: Introduction*). Not all broilers that die from PHS have ascites. Death may occur suddenly before clinical signs are seen.

Lesions

Most lesions are the result of increased venous hydraulic pressure secondary to RVF. There is a variable amount of clear yellow fluid and clots of fibrin in the hepatoperitoneal spaces. The liver may be swollen and congested, or firm and irregular with edema, and have clotted protein adherent to the surface. It may be nodular or shrunken; it may be white with subcapsular edema and a thickened capsule, or have large or small blebs of fluid between the capsule and the visceral peritoneum. Hydropericardium is mild to marked, and occasionally there is pericarditis with adhesions. Right ventricular dilatation and mild to marked hypertrophy of the right ventricular wall may be noted. The right atrium and vena cava are very dilated. Occasionally, there is thinning of the left ventricle. The lungs are extremely congested and edematous. The intestine may or may not be empty.

Diagnosis

Broilers that die from ascites or suddenly as the result of RVF or pulmonary hypertension can be identified by the enlarged heart; enlarged, thickened right ventricle; or fluid in the body cavities and heart sac. If the wall of the right ventricle is enlarged or thickened, the broiler has probably died from PHS, even if there is no fluid in the body or heart sac.

Control

Reducing the birds' metabolic oxygen requirement by slowing growth or reducing feed can prevent ascites caused by PHS. Environmental temperature, humidity, and air movement should be controlled to prevent excessive loss of body heat, particularly in the early neonatal period. Ascites caused by other factors (e.g. sodium, lung damage, liver damage, etc.) can be prevented by avoiding the etiologic agents involved. Altitudes >3,000 ft (900 m) are unsatisfactory for meat-type chickens, and growth must be slowed to prevent mortality. More care to prevent chilling is also necessary at higher altitudes.

Breast Blisters

In chickens and turkeys, a bursa lined with synovial membrane normally exists over the anterior projection of the keel bone. When this bursa becomes inflamed by trauma or infection, fluid or exudate accumulates and appears as a fluid-filled blister 1-3 cm in diametre. Causes of trauma to this bursa include poor feathering, hard flooring, and leg weakness, which is associated with increased recumbency.

Some young turkeys have pointed keels, which can lead to increased bursal trauma, but as the size of their breast muscle increases and trauma decreases, lesions may regress. Infectious causes of sternal bursitis include Mycoplasma synoviae, Staphylococcus, and Pasteurella spp.

Breast Buttons

These are lesions found in a similar location to breast blisters. They have a hard crust on the surface and a core of dead skin and granulomatous reaction extending into the subadjacent subcutis. Their etiology is not well defined but they are not due to the causes listed above for breast blisters. Rather they may be chemical burns due to prolonged contact of poorly feathered skin with wet litter containing ammonia or toxins.

Cannibalism

Cannibalism is an abnormal behaviour of chickens and turkeys most often manifested as vent-picking or picking at unfeathered skin on the head, comb, wattles, or toes. No single cause has been identified, but overcrowding, excessive light intensity, and nutritional imbalances are directly correlated with its occurrence. Additionally, in overly fat pullets entering egg production or hens in production, mucosa will protrude from the vent during and after egg laying, and this red tissue will attract pecking. Other factors that predispose to cannibalism are insufficient feeder space, mineral and vitamin deficiencies, skin injuries, and failure to remove dead birds daily. Other than the loss of birds due to pecking trauma, cannibalism often leads to transmission of infectious diseases (e.g. erysipelas) and botulism.

Control depends on correcting or reducing the above risk factors and trimming the sharp distal end of the upper beak to prevent pecking. Trimming of the tip of the beak distal to the nostrils is often done at 1 day of age and repeated between 6 and 12 wk of age in maturing pullets or turkeys. Cautery often is required to provide hemostasis.

Fluke Infections

Modern poultry production methods have diminished the incidence of fluke infections, although the parasites persist in poultry allowed contact with snails or other hosts and in some wild birds. Prosthogonimus macrorchis, the oviduct fluke of poultry, infects birds after they consume infective metacercariae in larval or mature dragonflies, the secondary host. The fluke matures in ~2 wk in the bursa of Fabricius or, in gallinaceous birds without a functional bursa (e.g., chickens, turkeys, pheasants), in the oviduct.

Light infections without clinical signs appear in ducks and other birds with a functional bursa. In gallinaceous birds, heavy infections in the oviduct cause inappetence, droopiness, weight loss, calcareous cloacal discharge, depressed egg production, and an increase in soft-shelled eggs. Lesions range from mild inflammation to distention or rupture of the oviduct; death may result. Diagnosis by faecal examination is unreliable because fluke eggs are not consistently present. Adult flukes may appear in the bird's eggs or be found in the oviduct on necropsy.

To prevent fluke transmission, birds must be kept from feeding on dragonflies. There is no effective treatment approved for use in poultry. Carbon tetrachloride, a common remedy, is highly toxic to chickens and other birds.

Collyriclum faba, another common fluke in birds, appear as subcutaneous cysts 4-6 mm in diametre (usually containing 2 adults) anywhere on the body but more frequently near the vent in turkeys, chickens, and other birds. The cysts ooze an exudate, which attracts flies and predisposes to bacterial infection. Signs in young birds include locomotor difficulty and inappetence; death may result in heavy infections. The parasites can be removed surgically. The life cycle is unknown but probably involves snails and insects such as dragonflies or mayflies. Prevention of infection requires restricting birds from areas frequented by aquatic insects.

Gout

Avian species excrete nitrogenous wastes as urates bound in colloidal form with mucus in their urine. Renal disease decreases the clearance of uric acid from the blood, which results in acute or chronic hyperuricemia, and the excess uric acid precipitates on either visceral or articular surfaces (gout). Urate deposits are white and semisolid and must be differentiated from yellow fibrinous or purulent inflammatory exudates that are secondary to infectious causes such as synovitis, peritonitis, perihepatitis, and pericarditis.

Acute urate deposition occurs after rapidly progressing renal failure, or as a terminal event with acute decompensation of chronic renal disease. Deposits develop most commonly on the pericardium, peritoneum, and liver capsule, and rarely on synovial surfaces of joints and tendons. They are usually present for too short a time to induce significant inflammation. Most clinical cases of acute renal failure and urate deposition in commercial poultry are due to dehydration, ingestion of feed containing >3% calcium by nonlaying chickens, renal infection by nephrotropic strains of infectious bronchitis virus, or infection with avian nephritis virus. Other avian species commonly develop visceral deposits secondary to nephrotoxin exposure, most commonly aminoglycoside antibiotics or heavy metals.

Chronic urate deposition is less common and occurs after longterm increases in serum levels of uric acid. Deposits develop on synovial membranes in the toes and wing joints and incite a chronic granulomatous reaction to urate crystals (tophi). Chronic urate deposition may be seen in chickens that have hereditary defects in uric acid metabolism or that are fed excessive protein. Urolithiasis is common in older laying chickens. Brittle, white, staghorn calcium urate calculi form in one or both ureters. Most cases are due to feeding high-calcium laying feed to hens not in egg production, infection with infectious bronchitis virus, or severe vitamin A deficiency. If blockage is complete, acute postrenal failure develops, and birds die with acute urate deposition on visceral surfaces or less commonly in joint spaces. If blockage is incomplete or unilateral, chickens survive in compensated renal failure, and chronic urate deposits form in joint spaces.

Pendulous Crop

Incidence of pendulous crop is low in flocks of chickens and turkeys. The crop is visibly distended and contains foul-smelling fluid, feed, and litter. Digestion is impaired, and affected birds become thin or emaciated. If these birds survive, they often are condemned or trimmed at processing to reduce contamination by ingesta. The etiology is not known, but a hereditary predisposition has been suggested in turkeys. Incidence may increase with erratic feed or water consumption. Experimentally feeding rations containing cerelose as a substitute for starch can cause pendulous crops. Vagus nerve damage has also been postulated as a cause. There is no known efficacious treatment.

Colibacillosis

Colisepticemia, Escherichia Coli Infection

Colibacillosis occurs as an acute fatal septicemia or subacute pericarditis and airsacculitis. It is a common systemic disease of economic importance in poultry and is seen worldwide.

Etiology and Pathogenesis

Escherichia coli is a gram-negative, rod-shaped bacterium normally found in the intestines of poultry and most other animals; although most are nonpathogenic, a limited number produce extraintestinal infections. Pathogenic strains are commonly of the O1, O2, and O78 serotypes, but serotypes O11, O15, O18, O51, O115, and O132 have also been reported for E coli isolates associated with cellulitis and colibacillosis. There is considerable diversity of serogroups among clinical isolates, and only a small percentage of these isolates belong to serotypes O1, O2, or O78.

In fact, 18-29% of avian E coli isolates cannot be typed. Therefore, no single E coli serotype used as a bacterin can provide full protection against all of the serotypes that cause E coli infections. Virulence factors include the ability to resist phagocytosis, utilisation of highly efficient iron acquisition systems, resistance to killing by serum, production of colicins, and adherence to respiratory epithelium. Virulent E coli are generally nontoxigenic, poorly invasive, and do not possess common adhesins.

Large numbers of E coli are maintained in the poultry house environment through faecal contamination. Initial exposure to pathogenic E coli may occur in the hatchery from infected or contaminated eggs, but systemic infection usually requires predisposing environmental factors or infectious causes. Mycoplasmosis, infectious bronchitis, Newcastle disease, hemorrhagic enteritis, and turkey bordetellosis precede colibacillosis. Poor air quality and other environmental stresses may also predispose to E coli infections.

Systemic infection occurs when large numbers of pathogenic E coli gain access to the bloodstream from the respiratory tract or intestine. Bacteremia progresses to septicemia and death, or the infection extends to serosal surfaces, pericardium, joints, and other organs.

Clinical Findings and Lesions

Signs are nonspecific and vary with age, organs involved, and concurrent disease. Young birds dying of acute septicemia have few lesions except for enlarged, hyperemic liver and spleen with increased fluid in body cavities. Birds that survive septicemia develop subacute fibrinopurulent airsacculitis, pericarditis, perihepatitis, and lymphocytic depletion of the bursa and thymus. (Unusually pathogenic salmonellae produce similar lesions in chicks.) Although airsacculitis is a classic lesion of colibacillosis, whether it results from primary respiratory exposure or from extension of serositis is unclear. Sporadic lesions include pneumonia, arthritis, osteomyelitis, and salpingitis.

Diagnosis

Unlike pathogenic E coli associated with illnesses in other animal species, avian isolates are generally nonhemolytic on sheep (5%) blood agar. Isolation of a pure culture of E coli from heart blood, liver, or typical visceral lesions in a fresh carcass indicates primary or secondary colibacillosis. Consideration should be given to predisposing infections and environmental factors. Pathogenicity of isolates is established when parenteral inoculation of young chicks or poults results in fatal septicemia or typical lesions within 3 days. Pathogenicity can also be detected by inoculation of the allantoic sac of 12 day old chick embryos. Resulting gross lesions include cranial and skin hemorrhages in addition to encephalomalacia in embryos inoculated with virulent isolates.

Treatment and Control

Treatment strategies include attempts to control predisposing infections or environmental factors and early use of antibacterials indicated by susceptibility tests. Most isolates are resistant to tetracyclines, streptomycin, and sulfa drugs, although therapeutic success can sometimes be achieved with tetracycline. In fact, 90% of clinical isolates are resistant to tetracycline, with 60% of isolates resistant to 5 or more antibiotics. Fluoroquinolone use is controversial because the use of these drugs in commercial broilers is believed to select for resistant Campylobacter spp associated with human foodborne infections. Commercial bacterins, administered to breeder hens or chicks, have provided some protection against homologous E coli serotypes.

Duck Viral Hepatitis

Duck viral hepatitis is an acute, highly contagious, viral disease of young ducklings characterised by a short incubation period, sudden onset, high mortality, and characteristic liver lesions. The disease is of economic importance in all duck-raising areas of the world. Three distinct types of duck hepatitis virus (DHV) have been isolated from diseased ducklings. A natural outbreak of DHV Type I has been reported in mallard ducklings; experimental DHV Type I infections have been produced in goslings, turkey poults, young pheasants, quail, and guinea fowl. The viruses that cause hepatitis in ducklings should not be confused with duck hepatitis B virus, a hepadnavirus infection of older ducks.

Etiology

The originally described, most widespread, and most virulent DHV Type I is an enterovirus in the family Picornaviridae and is readily propagated in chick and duck embryos. It does not produce hemagglutinins. Field experience with DHV Type I indicates that egg transmission does not occur. The disease can be transmitted experimentally by parenteral or oral administration of infected tissues.

Viruses differing from classic DHV Type I have been recognised as causes of hepatitis in ducklings. DHV Type II is considered to be an astrovirus and is difficult to propagate under laboratory conditions; DHV Type III is a member of the Picornaviridae, is antigenically distinct from Type I virus, and can be propagated in duck (but not chick) embryos. A distinct serologic variant of DHV Type I, named DHV Type Ia, has also been described.

Clinical Findings

The incubation period for Type I virus is 18-48 hr. Affected ducklings become lethargic, lose balance, paddle spasmodically, and die within minutes, typically with opisthotonos. Although adults may become infected, clinical signs have not been seen in ducks >7 wk old. Mortality may be as high as 95% in ducklings. Practically all deaths occur within 1 wk after onset of signs. The clinical course of DHV Type II infection is similar to that of Type I and can occur in ducklings immune to Type I infection. DHV Type III infections occur in ducklings despite immunity to Type I virus. The clinical course of Type III infection is less severe, and mortality is rarely >30%.

Lesions

The lesions caused by all 3 types of DHV are similar. The liver is enlarged and covered with hemorrhagic foci up to 1 cm in diametre. The spleen may be enlarged and mottled. Kidneys may be swollen, and renal blood vessels congested.

Diagnosis

A presumptive diagnosis can be based on the history and lesions. Sudden onset, rapid spread, and short course, together with characteristic liver lesions, are highly suggestive of duck viral hepatitis.

Type I virus may be isolated in duck embryos, day-old ducklings, and duck-embryo liver cell cultures, or less easily in chicken embryos. The virus can be identified by neutralisation with specific antisera or by inoculation into both susceptible and immune ducklings. Type II and III viruses are not neutralised by classic Type I antiserum.

Prevention and Treatment

Prevention is by strict isolation, particularly during the first 5 wk of age. Contact with wild waterfowl should be avoided. Rats have been reported as a reservoir host of the virus; therefore, pest control is indicated.

Immunisation of breeder ducks with modified live virus vaccines, using Type I, II, and III viruses, provides parenteral immunity that effectively prevents high losses in young ducklings. The Type I virus vaccine is administered SC in the neck to breeder ducks at 16, 20, and 24 wk of age and every 12 wk thereafter throughout the laying period. Three immunisations are advisable for passive protection of ducklings.

An inactivated DHV Type I vaccine for use in breeder ducks that have been previously primed with live DHV Type I virus has been described. A single dose of the inactivated vaccine, given IM before the birds come into lay, provides passive immunity for a complete laying cycle to progeny ducklings.

The chick-embryo origin, modified live Type I virus vaccine also can be used for early vaccination of ducklings susceptible to Type I (progeny of nonimmune breeders). This vaccine is administered SC or by foot web stab in a single dose to day-old ducklings. Vaccinated ducklings rapidly develop an active immunity over 3-4 days.

Antibody against Type I virus, prepared from the eggs of hyperimmunised chickens, administered SC in the neck at the time of initial loss, is an effective flock treatment.

Enterococcosis

The application of new bacteriologic techniques, especially DNA-DNA and DNA-rRNA hybridisation has led to the reclassification of Lancefield group D streptococci as Enterococcus spp. (For a discussion of diseases caused by the Lancefield antigenic serogroup C and other Streptococci spp). Enterococcus spp in avian species are worldwide in distribution. Enterococci are ubiquitous in nature and commonly found in various poultry environments. Enterococcus spp are considered normal microflora of the intestinal tract of poultry and other birds. A high percentage of ready-to-eat poultry products are contaminated with Enterococcus spp; however, no resultant food poisoning in humans has been reported.

Etiology and Epidemiology

The genus Enterococcus is composed of gram-positive, spherical bacteria occurring singly, in pairs, or in short chains, which are nonmotile, non- sporeforming, facultative anaerobes. They are catalase-negative and ferment sugars, usually to lactic acid. Common avian isolates can be differentiated by their ability to ferment mannitol, sorbitol, and L-arabinose and by their growth on MacConkey agar without crystal violet or salt. (Other types of MacConkey agar inhibit Enterococcus and may provide false-negative results.) Enterococcus spp isolated from avian species and associated with disease include E faecalis, E faecium, E durans, E avium, and E hirae. E faecalis affects birds of all ages; it is a serious disease occurring in embryos and young chicks from faecal-contaminated eggs. E faecium is a cause of mortality in ducklings.

Enterococci are transmitted most commonly via oral and aerosol routes. However, transmission can occur through skin injuries, especially in caged layers. Aerosol transmission of E faecalis results in acute septicemia in chickens. Concurrent enteric infections or any condition compromising the intestinal villous epithelium, allowing penetration of resident enterococci, can result in septicemia, bacterial endocarditis, or both. Incubation periods range from 1 day to several weeks, with 5-21 days most common. Endocarditis can occur when a septicemic enterococcal infection progresses to a subacute or chronic stage. Enterococcus spp have been associated with brain necrosis and encephalomalacia in young chickens. Some enterococci, however, have been demonstrated to have a beneficial effect on growth and feed efficiency and are being studied as potential probiotics.

Clinical Findings

Enterococcus spp in poultry can result in 2 distinct clinical forms of disease, acute and subacute/chronic. In the acute form, clinical signs are related to septicemia and include depression, lethargy, lassitude, pale combs and wattles, ruffled feathers, diarrhea, fine head tremors, and decrease or cessation of egg production. Often, only dead birds are found. In the subacute/chronic form, depression, loss of body weight, lameness, and head tremors may be observed. Body temperature is elevated in birds with persistent bacteremia. Clinically affected birds eventually die if not treated. Egg transmission or faecal contamination of hatching eggs results in late embryo mortality and an increased number of chicks or poults unable to "pip" or penetrate through the shell at hatch.

Lesions

Gross lesions of enterococci infection in acute disease include splenomegaly, hepatomegaly (with or without foci), enlarged kidneys, and congestion of subcutaneous tissue. Omphalitis or enlarged yolk sacs may be seen in chicks or poults infected at hatching. Hepatomegaly, splenic necrosis, fibrinous pericarditis, perihepatitis, and airsacculitis are observed in ducks infected with E faecium. Lesions of chronic enterococcal infections include fibrinous arthritis and/or tenosynovitis, osteomyelitis, fibrinous pericarditis and perihepatitis, necrotic myocarditis, and valvular vegetative endocarditis similar to that observed with Streptococcus zooepidemicus infection. Additional gross lesions associated with valvular endocarditis include an enlarged, pale, flaccid heart; pale to hemorrhagic areas in the myocardium; infarcts in the liver, spleen, or heart; and, less commonly, infarcts in the lung, kidney, and brain.

On microscopic examination, the liver has dilated sinusoids congested with RBC and increased heterophils. Splenomegaly is characterised by congestion and hyperplasia of cells in the mononuclear phagocytic system. Valvular lesions consist primarily of fibrin with bacteria, heterophils, macrophages, and fibroblasts. Other microscopic lesions related to endocarditis include cerebral vasculitis and infarcts, leptomeningitis, glomerulonephritis, and thrombosed pulmonary vessels. Focal granulomas can be found in virtually any tissue as a result of septic emboli. Aggregates of bacteria are present throughout necrotic areas with a zone of heterophils just within the necrotic border, a characteristic feature of the lesion. Gram-positive bacterial colonies are readily observed in thrombosed vessels and within necrotic foci.

Diagnosis

Demonstration of bacteria typical of enterococci in blood or impression smears of affected heart valves or lesions from birds with typical clinical signs will provide a presumptive diagnosis of enterococcosis. Isolation of Enterococcus spp (without faecal contamination) from typical lesions will confirm the diagnosis.

Enterococci are easily isolated on blood agar or more specific differential media, which should help differentiate species. Fermentation of mannitol, sorbitol, and arabinose, and growth on MacConkey agar (without crystal violet or salt) can also aid in differentiating enterococci from Streptococcus spp. Preferred tissues for culture include liver, spleen, blood, yolk, embryo fluids, or any suspected lesion. Diagnosis of bacterial endocarditis is based on valvular vegetations with secondary infarcts of myocardium, liver, or spleen. In suspected cases, it is important to culture lesions to establish a definitive diagnosis and rule out other bacteria.

Differential diagnosis includes other bacterial septicemic diseases, e.g., staphylococcosis, streptococcosis, colibacillosis, pasteurellosis, and erysipelas.

Treatment and Prevention

Treatment includes use of antibiotics such as penicillin, erythromycin, novobiocin, oxytetracycline, chlortetracycline, or tetracycline in acute and subacute infections. Clinically affected birds respond well early in the course of the disease. As the disease progresses within a flock, treatment efficacy decreases. Antibacterial sensitivity should be performed on bacterial isolates in any clinical cases of enterococcosis before treatment begins. There is no treatment for poultry with bacterial endocarditis.

Prevention and control requires reducing stress and preventing immunosuppressive diseases and conditions. Proper cleaning and disinfection can reduce environmental enterococcal resident flora to minimise external exposure.

Erysipelas

Erysipelas in poultry is seen worldwide, mainly as an acute septicemia. Outbreaks usually occur suddenly, with a few birds being found dead followed by increasing mortality on subsequent days. Mortality may range from <1% to 50%. From an economic standpoint, turkeys are the most important poultry species affected, but serious outbreaks have occurred in chickens, ducks, and geese. Mammals are also affected, with swine being the most economically important species. Infection in reptiles and amphibians has also been reported. The organism has been isolated from the surface slime on fish, which may serve as a source of infection for other species. People usually become infected when the organism enters through cuts in the skin. There have been no reports of people becoming infected by the oral route. The disease in humans (erysipeloid) is most common in people who handle infected tissues such as veterinarians, butchers, and fish handlers. Erysipeloid in people may be a localised or a septicemic and occasionally fatal infection.

Etiology

The causative agent is Erysipelothrix rhusiopathiae, a facultatively anaerobic, intracellular bacterium. A second genomic species, E tonsillarum, has been described but is not considered pathogenic for poultry. Morphologically, E tonsillarum cannot be distinguished from E rhusiopathiae. E rhusiopathiae stains gram-positive but tends to decolourise, particularly in older cultures. The organism is nonmotile, does not form spores, and produces no known toxins. There is no flagellum but a capsule has been demonstrated. The cellular morphology of E rhusiopathiae is variable. Cells freshly isolated from tissues during acute infection or from smooth colonies are straight or slightly curved small rods that may occur in short chains. Cells from older cultures or rough colonies tend to become filamentous and may be confused with mycelia. The filamentous form occurs more frequently after repeated passages on artificial media.

E rhusiopathiae grows readily on ordinary culture media containing the blood or sera of various animals. Growth is enhanced by reducing the oxygen content or increasing the carbon dioxide level to 5-10%. Optimal incubation temperature is 35-37°C, and the optimal pH range is 7.4-7.8.

The organism is not readily destroyed by the usual laboratory disinfectants, and it may survive in litter or soil for various lengths of time; therefore, disinfection of premises is difficult. E rhusiopathiae may also survive smoking and pickling processes. It is inactivated by a 1:1,000 concentration of bichloride of mercury, 0.5% sodium hydroxide solution, 3.5% liquid cresol, 5% solution of phenol, or 0.5% formalin.

Though different serotypes of E rhusiopathiae exist, no correlation has been shown to exist between the serotype, chemical structure, or biochemical pattern, and the manifestation of the septicemic, urticarial, or endocardial forms of erysipelas.

Epidemiology

Erysipelas occurs sporadically in poultry of all ages. Turkeys are susceptible regardless of sex or age. Recent evidence indicates that there may be a genetically related resistance in turkeys. The incidence in males is reported to be higher, but this is not supported by experimental data. Erysipelas may affect the fertility of males and may contribute to downgrading and processing losses. Infection results from entrance of the organisms through breaks in the skin, through the mucous membranes such as during artificial insemination, by ingestion of contaminated foodstuffs (particularly cannibalism of infected carcasses), and possibly by mechanical transmission via biting insects. Fighting and cannibalism increase losses.

The organism is shed in feces from infected animals and contaminates the soil, in which it may survive for long periods depending on temperature and pH. Seasonal changes in climate such as the onset of cold, rainy weather have been associated with disease occurence. Poultry, as well as other animals, may be carriers and shed the organism without showing clinical signs of disease. In nonvaccinated flocks, morbidity and mortality may reach 40-50%, but mortality is usually limited to <15%. In vaccinated flocks, some birds may be depressed for a short period and recover. Mortality in vaccinated and nonvaccinated poultry is influenced by the virulence of the organism.

Clinical Findings

Erysipelas is primarily an acute infection that results in sudden death. In an affected flock, a few birds may be depressed but easily aroused; within 24 hr, a few birds will be dead. Just before death, some birds may be very droopy, with an unsteady gait. Chronic clinical disease in a flock is not usual but does occur; birds may have cutaneous lesions and swollen hocks. Turkeys with vegetative endocarditis usually do not have clinical signs and may die suddenly. Erysipelas should be suspected in flocks that have been artificially inseminated 4-5 days before an episode of death without clinical signs. Clinical signs in chickens include general weakness, depression, diarrhea, and sudden death. In laying hens, egg production may drop markedly.

Lesions

At necropsy, a generalised darkening of the skin or various sized areas of diffuse darkening is common. The liver and spleen are usually enlarged and friable and may be mottled. There may be other gross lesions such as peritonitis, pericarditis, catarrhal exudate in the GI tract, and degeneration of fat associated with the thigh and heart.

11

Epidemic Threat from Poultry Farming

Bird Flu is a Real Pandemic Threat to Humans

As their lungs filled ... the patients became short of breath and increasingly cyanotic. After gasping for several hours they became delirious and incontinent, and many died struggling to clear their airways of a blood-tinged froth that sometimes gushed from their nose and mouth. It was a dreadful business.

Isaac Starr, 3rd year medical student, University of Pennsylvania, 1918. All the genes of all influenza viruses in the world are being maintained in aquatic birds, and periodically they transmit to other species... The 1918 viruses are still being maintained in the bird reservoir. So even though these viruses are very ancient, they still have the capacity to evolve, to acquire new genes, new hosts. The potential is still there for the catastrophe of 1918 to happen again.

Dr. Robert Webster, Influenza Expert, Present Day.

The pair of quotes presented above were used to begin and end a short essay I wrote many years ago on The 1918 Spanish Flu Pandemic. Isaac Starr's words offer powerful testimonial to the harsh lethality of a killer stain of influenza, and Dr. Robert Websters observation reminds us that it would be dangerous to regard the events of 1918 as an isolated slice of history that we can safely file away and forget.

In the early part of 1995, as I cast around in search of the disease that would ultimately prove to be the true villian of the science thriller I was planning to write, influenza was scarcely on the radar scope of anyone but virologists specialising in the field of emerging diseases.

Because influenza exists in the human population in an ever-present but relatively benign form, changing just a little from year to year, we tend, quite mistakenly, to think we understand the nature of the threat that it presents.

Bird Flu cannot be eradicated, we must learn to live with it.

Despite the apparent "novelty" of a strain of bird flu that can sweep across the globe and kill not only the domesticated foul that carries the virus causing the pandemic, but also those humans unfortunate enough to get too close to the sick birds they tend, finding influenza in domesticated birds is not the least bit uncommon.

The natural reservoir for influenza virus is in fact the intestines of water fowl. In particular, ducks habour the virus, and suffer no ill effects from doing so. Because these migratory birds are always on the move, and because they shed virus in the waterways that they inhabit, influenza virus travels widely and is easily transmitted from bird to bird.

Because of this, domesticated birds, like the turkey and the chicken, which cross paths with the virus, become infected and habor the potential to transmit the virus to humans (for example, virus secreted onto the surface of eggs could end up infecting egg handlers—for this reason, if bird handlers notice any of the following symptons in their birds they should immediately notify Federal or State animal health officials: sudden death; lack of energy and appetite; decreased egg production; soft-shelled or misshapen eggs; swelling; purple discoloration; nasal discharge; coughing, sneezing; lack of coordination and diarrhea).

Often an intermediate host, the pig, is involved in the cross-species transmission of influenza. When this happens, the result is referred to as Swine Flu. Even so, the source of the virus is generally aquatic birds.

Most strains of bird flu *do not* cross species and end up infecting humans. However the H5N1 strain that has in the last decade been known to infect humans, and kill them in 5 out of every 10 cases, clearly has surmounted the first of two barriers that must be cleared before a bird flu pandemic can take hold in the human population. The second barrier will be surmounted when this strain acquires the ability to jump from one human to another. This may never happen, and hopefully will not. But based on the history of pandemic flu, it seems more likely that this is just a matter of time. Whether this takes one year or twenty is just impossible to say. The World Health Organisation has estimated that a human-to-human strain of bird flu could result in an influenza pandemic with the potential to cause between 2 and 7 million fatalities worldwide. Considering the underwhelming state of health services offered by the majority of countries in the world today, these numbers do not appear to be overhyped estimates of the potential lethality of a bird flu pandemic. Should a bird flu pandemic break out with a lethality approaching that of the 1918 Spanish Flu pandemic, it is probably fair to state that the WHO prediction is likely to fall far short of the actual death toll.

Emergence of a Bird Flu Pandemic?

The current candidate for the cause of a bird flu pandemic that crosses into the human population is the H5N1 strain. As discussed in my essay on The 1918 Spanish Flu Pandemic, and the Emerging Bird Flu Pandemic, this strain was first detected in Hong Kong in 1997, and in the time since then has spread halfway across the globe despite strenuous efforts to eradicate it in poultry stocks that have become infected. This movement of the virus, which requires transmission from one (bird) host to another via shared waterways, is slow in comparison to the movement of an airborne strain which can be transmitted from one human host to another in the particulate matter exchanged during coughing and sneezing events. The pandemic of 1918 spread across the globe in about six months, despite the dependence of that time on ship borne travel. In an age of international travel, this encircling time could be reduced to a matter of days. Once it begins, the likelihood that it can be stopped in its tracks is small.

Life during a Bird Flu Pandemic

One can never predict with any accuracy the way a natural catastrophe will unfold. But some speculation on life during the time of a bird flu pandemic seems worthwhile. Things we can say for certain are the following:

No country will be fully prepared for the event. This stems from the fact that no government is ever prepared for any natural catastrophe, as such preparations are prohibitively expensive. This means the bulk of the population will not have access to medical treatment when they need it. Families will be required to look after themselves. Only the patients identified in the very first stages of a bird flu pandemic are likely to receive treatment.

Medical supplies for combating a bird flu pandemic in the human population will be quickly exhausted. Even now, access to such antivirals as Tamiflu is poor in most parts of the world. (In late 2005 the makers of Tamiflu stepped up production to 10 times the normal level in response to worldwide demand for the drug in anticipation of an impending pandemic).

Likewise, flu vaccines, which require a long lead time for preparation (six months or more) and can easily be "spoiled" due to inadequate preparation, will not be available in anything like a sufficient supply in the foreseeable future.

Social disruption will be significant. When workers are faced with the decision of weighing the safety of family members against the worth of their job, you can bet the absentee rate will skyrocket during a pandemic.

When the loss of workers begins to affect the delivery of food supplies, power, and water, we are suddenly looking at a different kind of emergency scenario—one reminiscent of the scenes that played out in New Orleans after Hurricane Katrina struck in 2005. When the services we take for granted every day suddenly disappear, people become desparate and things go to hell rather quickly. The break down of social order may be the most frightening prospect that an emerging bird flu pandemic could bring about.

Preparing for a Bird Flu Pandemic

It is tempting for all of us, the author included, to stick our heads in the sand and hope that the news stories about the possibly of an impending *bird flu pandemic* are mostly hype. But this is about as wise as living in Florida and hoping you have seen the last hurricane, or living in Los Angeles and assuming that you'll not be hit by another earthquake. Really it is just a matter of time, though no one knows how much.

There are some not so difficult things you can do to prepare, things that will also help prepare you for other emergency events. These seems practical to me:

Stock up on water: You can never have too much of this in storage, and it is cheap. Have at least a 3 month supply in stock at any time (6 months if practical), and rotate your supply so that you use up the oldest bottles at the rate you replace them with new ones. This way you will never have to worry about filtering existing water supplies if it comes to water shortages in your area.

Stock up on food supplies: This one is a little more difficult, but if you are lucky enough to have one of those huge freezers in your home, fill it up.

Rotate your food, so that it is never more than 3 (or 6) months old. Be sure to buy only the foods you normally buy, as you'll be eating them regardless of whether or not any disaster strikes.

Dry foods can be stored separately, but people tend too forget about them and only check their supplies in the advent of an emergency. So be prepared to toss these after a year when they are no longer edible.

Purchase some quantity of air filtration masks which, in the event of a true bird flu pandemic will be regarded as indispensible when you need to leave the house and mix with the general populace.

As regards what type of mask you should obtain if you are inclined to purchase a box or two (not a bad idea as you won't be able to do it if the time comes that you need to wear them) the Centres for Disease Control and Prevention advise health workers to use a "fit-tested respirator, at least as protective as a National Institute of Occupational Safety and Health (NIOSH)-approved N-95 filtering facepiece (i.e., disposable) respirator" as one of the possible precautions against airborne infection when working with patients with known or suspected SARS or Bird Flu. 3M manufactures such masks, and the latest variety with the "Cool Flow" exhalation valve for increased comfort during extended wearing periods seems to be the mask of choice.

For other possible preparedness items you might check out a Distributor of Flu Preparedness Items (naturally enough!).

If it comes down to it, and an infectious form of bird flu arrives in your country, one of the most important things you can do to avoid infection is practice safe hygiene. This means that you go into obsessive compulsive mode and assume that everything you touch has the potential to be carrying virus. It's somewhat pointless to assume this of the contents of your own home (unless you have a stricken person on site) but once you leave the house, it is a good operating assumption. This means under no circumstances do you touch your hands to your face while in public.

Children, and adults, who have the habit of putting their fingers in their eyes and noses are *extremely susceptible* to infection in such an environment.

When you return to your home, clothing goes into the laundry immediately and you wash your hands thoroughly with soap to rinse away any virus you might have come in contact with. You always use a face mask in public if you have access to one. Otherwise you make one. Common sense suggests that anything that inhibits the likelihood of inhaled aerosolised influenza is better than nothing. But a proper face mask is designed to be efficient at this task. These are designed to be used *once only* and discarded.

If you are a vendor, think about how you might change the operational nature of your business to stay afloat during a time of crisis. What can you do to reduce the likelihood that your employees will desert you? For instance, if you sell clothing you are very unlikely to receive my business if I must enter an enclosed space (a building) in order to purchase something.

At such a time I do not want to breathe the air of a confined space. But if your stock was accessible in an open-air market I would probably be much more inclined. Employees will feel the same way, particularly if the goods in question are not critical to day to day survival.

Finally, do some more reading on the subject of bird flu pandemic. The information about bird flu found on this page is by no means exhaustive. Reading resources are listed below in the Related Links section.

Could a Bird Flu Pandemic resemble the Spanish Flu Pandemic of 1918?

A lot has happened in the last decade as regards our scientific understanding of the genetic mechanisms responsible for the virulence of some strains of influenza. Much of this new understanding has been a direct result of the work of Dr. Jeffery Taubenberger and his team at the Armed Forces Institute of Pathology.

Back in 1998 I interviewed Dr. Taubenberger about his effort to reconstruct the genes of the 1918 Spanish Flu virus. At the time I had forseen the day when it would be possible to reconstruct the entire virus.

In October of 2005, after 10 years of work by Taubenberger's team, it was finally reported that the virus had been re-engineered. You can read about the announcement in the PBS article entitled 1918 Spanish Flu Offers Clues About Pandemic Viruses. Of particular interest is the observation that Taubenberger's work "published in the journal Nature in October 2005, offered some parallels between the

Spanish flu virus and the H5N1 strain of the bird flu slowly spreading through Asia and recently turning up in other parts of the world." The implication here is clear. If the currently circulating H5N1 bird flu virus does acquire human-to-human transmission, the estimated mortality rates offered by the World Health Organisation for a bird flu pandemic could prove to be wildy optimistic.

Influenza Vaccine

The influenza vaccine, also known as flu shot, is an annual vaccine to protect against the highly variable influenza virus. Each injected seasonal influenza vaccine contains three influenza viruses: one influenza type A subtype H3N2 virus strain, one influenza type A subtype H1N1 (seasonal) virus strain, and one influenza type B virus strain.

Purpose and Benefits of Annual Flu Vaccination

"Influenza vaccination is the most effective method for preventing influenza virus infection and its potentially severe complications."

Deadly Epidemics each Winter

An influenza epidemic emerges during flu season each winter. There are two flu seasons annually, corresponding to the occurrence of winter in opposite months in the Northern and Southern Hemispheres.

Worldwide, seasonal influenza kills an estimated 250,000 to 500,000 people each year. Tens of thousands of Americans die in a typical flu season, but there are notable variations from year to year. In 2010 the Centres for Disease Control and Prevention (CDC) in the United States changed the way it reports the 30 year estimates for deaths from influenza. Now they are reported as a range from a low of about 3,300 deaths to a high of 49,000 per year over the past 30 years. The majority of deaths in the industrialised world occur in adults aged 65 and over. A review at the NIAID division of the NIH in 2008 concluded that "Seasonal influenza causes more than 200,000 hospita-lisations and 41,000 deaths in the U.S. every year, and is the seventh leading cause of death in the U.S." The average total economic costs caused by the annual influenza outbreak in the U.S. have been estimated at over $80 billion.

The number of annual influenza-related hospitalisations is many times the number of deaths. "The high costs of hospitalising young children for influenza creates a significant economic burden in the United States, underscoring the importance of preventive flu shots for children and the people with whom they have regular contact..." In 2006 the United States began recommending influenza vaccinations for preschoolers but Canada did not follow suit until 2010, "thereby creating a natural experiment to evaluate the effect of the policy in the United States."

A Canadian study found emergency room visits significantly lower for 2- to 4 year olds in Boston than in Montreal through the period (34% fewer ER trips).

Vaccination of preschoolers may have reduced their likelihood of transmission of flu to older siblings and raised the chances that their parents would vaccinate older children as well, since there were also 18 percent fewer emergency room visits by 5- to 18 year olds in Boston than Montreal during the study period.

In another six-year observational study, vaccination of children aged 6 months through 5 years was found to prevent illness in more than half.

National Advice on Flu Vaccination

In Canada, the National Advisory Committee on Immunisation, the group that advises the Public Health Agency of Canada, currently recommends that everyone aged 2 to 64 years be encouraged to receive annual influenza vaccination, and that children between the age of six and 24 months, and their household contacts, should be considered a high priority for the flu vaccine.

In the United States, "Routine influenza vaccination is recommended for all persons aged e"6 months." The sole group for whom routine influenza vaccination is still not recommended is infants less than six months of age. Within its blanket recommendation for general vaccination, the United States, the Centres for Disease Control and Prevention (CDC) emphasizes to clinicians the special urgency of vaccination for members of certain vulnerable groups, and their caregivers:

Vaccination is especially important for people at higher risk of serious influenza complications or people who live with or care for people at higher risk for serious complications.

Benefits of Vaccination

Vaccination against influenza is also, according to research published in July 2010, thought to be important for members of high-risk groups who would be likely to suffer complications from influenza, for example pregnant women and children and teenagers from six months to 18 years of age; In expanding the new upper age limit to 18 years, the aim is to reduce both the time children and parents lose from visits to pediatricians and missing school and the need for antibiotics for complications.

An added expected benefit would be indirect — to reduce the number of influenza cases among parents and other household members, and possibly spread to the general community.

Vaccination of school-age children has a strong protective effect on the adults and elderly with whom the children are in contact. Children born to mothers who received flu vaccination while pregnant are strongly protected from having to be hospitalised with the flu. "The effectiveness of influenza vaccine given to mothers during pregnancy in preventing hospitalisation among their infants, adjusted for potential confounders, was 91.5%."

Healthy, working adults who received influenza vaccine reported 25 percent fewer episodes of upper respiratory illness than those who received the placebo (105 vs. 140 episodes per 100 subjects, $P < 0.001$), 43 percent fewer days of sick leave from work due to upper respiratory illness (70 vs. 122 days per 100 subjects, $P = 0.001$), and 44 percent fewer visits to physicians' offices for upper respiratory illnesses (31 vs. 55 visits per 100 subjects, $P = 0.004$). The study, reported in the NEJM, estimated cost savings at $46.85 per person vaccinated, and concluded that "Vaccination against influenza has substantial health-related and economic benefits for healthy, working adults."

Influenza vaccination has been shown highly effective in health care workers, with minimal adverse effects. In a study of forty matched nursing homes, staff influenza vaccination rates were 69.9% in the vaccination arm versus 31.8% in the control arm. The vaccinated staff experienced a 42% reduction in sick leave from work ($P=.03$). A review of eighteen studies likewise found a strong net benefit to health care workers. The two of these eighteen studies that assessed the relationship of patient mortality relative to staff influenza vaccine uptake both found that higher rates of health care worker vaccination correlated with reduced patient deaths. An analysis of data and patient population health in New Mexico's 75 long-term care facilities nursing homes found that as vaccination rates of health care personnel with direct patient contact rose from 51 to 75 percent, the chances of a flu outbreak among patients in that facility went down by 87 percent. The New Mexico study showed that vaccinating health care personnel provided more protection to residents than vaccinating residents themselves.

In a 2010 survey of healthcare workers, 63.5% reported that they received the flu vaccine during the 2010-11 season, an increase from 61.9% reported the previous season. Health professionals with direct patient contact had higher vaccination uptake, such as physicians and dentists (84.2%) and nurse practitioners (82.6%).

Cross-protection

Annual seasonal flu vaccination may provide some level of protection against novel flu viruses. A number of studies suggest that seasonal flu vaccine may offer cross-protection, both against the H5N1-type (avian influenza) H5N1 infection and the 2009 flu pandemic (the H1N1 "swine flu.") Vaccine protection can be long-lasting. Participants who received vaccination against the swine flu in 1976 still enjoyed benefits 33 years later, exhibiting a significantly enhanced immune response to the 2009 pandemic H1N1.

History of the Flu Vaccine

Vaccines are used in both humans and nonhumans. Human vaccine is meant unless specifically identified as a veterinary, poultry or livestock vaccine.

Influenza

The first influenza pandemic was recorded in 1580; since this time, various methods have been employed to eradicate its cause. The etiological cause of influenza, the orthomyxoviridae was finally discovered by the Medical Research Council (MRC) of the United Kingdom in 1933.

Known Flu Pandemics

- 1889–90 — Asiatic (Russian) Flu, mortality rate said to be 0.75–1 death per 1000 possibly H2N2.
- 1900 — Possibly H3N8.
- 1918–20 – Spanish Flu, 500 million ill, at least 20–40 million died of H1N1.
- 1957–58 – Asian Flu, 1 to 1.5 million died of H2N2.
- 1968–69 – Hong Kong Flu, 3/4 to 1 million died of H3N2.
- 2009 - Swine Flu, caused by H1N1/09, 14,286 died.

Flu Vaccine Origins and Development

In the world wide Spanish flu pandemic of 1918, "Physicians tried everything they knew, everything they had ever heard of, from the ancient art of bleeding patients, to administering oxygen, to developing new vaccines and sera (chiefly against what we now call *Hemophilus influenzae*—a name derived from the fact that it was originally considered the etiological agent—and several types of pneumococci). Only one therapeutic measure, transfusing blood from recovered patients to new victims, showed any hint of success."

In 1931, viral growth in embryonated hens' eggs was reported by Ernest William Goodpasture and colleagues at Vanderbilt University. The work was extended to growth of influenza virus by several workers, including Thomas Frances, Wilson Smith and Macfarlane Burnet, leading to the first experimental influenza vaccines.

In the 1940s, the US military developed the first approved inactivated vaccines for influenza, which were used in the Second World War. Greater advances were made in vaccinology and immunology, and vaccines became safer and mass-produced. Today, thanks to the advances of molecular technology, we are on the verge of making influenza vaccines through the genetic manipulation of influenza genes.

Flu Vaccine Acceptance

According to the CDC: "Influenza vaccination is the primary method for preventing influenza and its severe complications. Vaccination is associated with reductions in influenza-related respiratory illness and physician visits among all age groups, hospitalisation and death among persons at high risk, otitis media among children, and work absenteeism among adults. Although influenza vaccination levels increased substantially during the 1990s, further improvements in vaccine coverage levels are needed".

The current egg-based technology for producing influenza vaccine was created in the 1950s. In the U.S. swine flu scare of 1976, President Gerald Ford was confronted with a potential swine flu pandemic. The vaccination program was rushed, yet plagued by delays and public relations problems. Meanwhile, maximum military containment efforts succeeded unexpectedly in confining the new strain to the single army base where it had originated. On that base a number of soldiers fell severely ill, but only one died. The program was cancelled, after about 24% of the population had received vaccinations. An excess in deaths of twenty-five over normal annual levels as well as 400 excess hospitalisations, both from Guillain-Barré syndrome, were estimated to have occurred from the vaccination program itself, illustrating that vaccine itself is not free of risks.

The result has been cited to stoke lingering doubts about vaccination. In the end, however, even the maligned 1976 vaccine may have saved lives. A 2010 study found a significantly enhanced immune response against the 2009 pandemic H1N1 in study participants who had received vaccination against the swine flu in 1976.

Current Status

Influenza research includes molecular virology, molecular evolution, pathogenesis, host immune responses, genomics, and epidemiology. These help in developing influenza countermeasures such as vaccines, therapies and diagnostic tools.

Improved influenza countermeasures require basic research on how viruses enter cells, replicate, mutate, evolve into new strains and induce an immune response. The Influenza Genome Sequencing Project is creating a library of influenza sequences that will help us understand what makes one strain more lethal than another, what genetic determinants most affect immunogenicity, and how the virus evolves over time. Solutions to limitations in current vaccine methods are being researched.

The rapid development, production, and distribution of pandemic influenza vaccines could potentially save millions of lives during an influenza pandemic. Due to the short time frame between identification of a pandemic strain and need for vaccination, researchers are looking at novel technologies for vaccine production that could provide better "real-time" access and be produced more affordably, thereby increasing access for people living in low- and moderate-income countries, where an influenza pandemic may likely originate, such as live attenuated (egg-based or cell-based) technology and recombinant technologies (proteins and virus-like particles).

As of July 2009, more than 70 known clinical trials have been completed or are ongoing for pandemic influenza vaccines. In September 2009, the US Food and Drug Administration approved four vaccines against the 2009 H1N1 influenza virus (the current pandemic strain), and expected the initial vaccine lots to be available within the following month.

Prospects for Universal Flu Vaccines

Many groups world wide are working on a universal flu vaccine that will not need changing each year, as the sector has been viewed as "increasingly hot". Companies pursuing the vaccine as of 2009 and

2010 include BiondVax, Theraclone, Dynavax Technologies Corporation, VaxInnate, Crucell NV, Inovio Pharmaceuticals, and Immune Targeting Systems (ITS).

In 2008 Acambis announced work on a universal flu vaccine (ACAM-FLU-ATM) based on the less variable M2 protein component of the flu virus shell.

In 2009, the Wistar Institute received a patent for using "a variety of peptides" in a flu vaccine, and announced it was seeking a corporate partner.

In 2010, the National Institute of Allergy and Infectious Diseases (NIAID) of the U.S. NIH announced a breakthrough; the effort targets the stem, which mutates less often than the head of the virus.

DNA vaccines such as VGX-3400X (aimed at multiple H5N1 strains) contain DNA fragments (plasmids). Inovios SynCon DNA vaccines include H5N1 and H1N1 subtypes.

In July 2011, F16 researchers created an antibody, which targets a protein found on the surface of all influenza A viruses called haemagglutinin.

F16 is the only known antibody that treats all 16 subtypes of the influenza A virus and might be the lynchpin for a universal influenza vaccine.

Other Vaccines are Polypeptide Based

Some universal flu vaccines have started early stage clinical trials.

- BiondVax are targeting the less variable stalk of the haemagglutinin molecule with Multimeric-001. This is aimed at type A (inc H1N1) and Type B influenza and has started a phase IIa study.
- Dynavax have developed a vaccine N8295 based on two highly conserved antigens NP and M2e and their TLR9 agonist, and started clinical trials in June 2010.
- ITS's fp01 includes 6 peptide antigens to highly conserved segments of the PA, PB1, PB2, NP & M1 proteins, and has started phase I trials.

Based on the results of animal studies, a universal flu vaccine may use a two-step vaccination strategy — priming with a DNA-based HA vaccine followed by a second dose with an inactivated, attenuated, or adenovirus-vector–based vaccine.

Some people given a 2009 H1N1 flu vaccine have developed broadly protective antibodies which has raised hopes for a universal flu vaccine.

Clinical Trials of Vaccines

A vaccine is assessed by the reduction of the risk of disease that is produced by vaccination, the vaccine's *efficacy*. In contrast, in the field, the *effectiveness* of a vaccine is the practical reduction in risk for an individual when they are vaccinated under real-world conditions.

Measuring efficacy of influenza vaccines is relatively simple, as the immune response produced by the vaccine can be assessed in animal models, or the amount of antibody produced in vaccinated people can be measured, or most rigorously, by immunising adult volunteers and then challenging with virulent influenza virus. In studies such as these, influenza vaccines showed high efficacy and produced a protective immune response.

For ethical reasons, such challenge studies cannot be performed in the population most at risk from influenza – the elderly and young children. However, studies on the effectiveness of flu vaccines in the real world are uniquely difficult. The vaccine may not be matched to the viruses in circulation that year; virus prevalence varies widely between years, and influenza is often confused with other influenza-like illnesses.

Nevertheless, multiple clinical trials of both live and inactivated influenza vaccines against seasonal influenza have been performed and their results pooled and analysed in several recent meta-analyses. Studies on live vaccines have very limited data, but these preparations may be more effective than inactivated vaccines. The meta-analyses examined the efficacy and effectiveness of inactivated vaccines against seasonal influenza in adults, children, and the elderly. In adults, vaccines show high efficacy against the targeted strains, but low effectiveness overall, so the benefits of vaccination are small, with a one-quarter reduction in risk of contracting influenza but no significant effect on the rate of hospitalisation.

However, the risk of serious complications from influenza is small in adults, so unless the effect from vaccination is large it might not have been detected. In children, vaccines again showed high efficacy, but low effectiveness in preventing "flu-like illness". In children under two the data are extremely limited, but vaccination appeared to confer no measurable benefit.

In the elderly, vaccination does not reduce the frequency of influenza, but seems to reduce pneumonia, hospital admission and deaths from influenza or pneumonia. However, the current data on the effectiveness of influenza vaccines in the elderly may be unreliable, due to high levels of selection bias.

Overall, the benefit of influenza vaccination is clear in the elderly and vaccination of children may be beneficial. Routine vaccination of adults is not predicted to produce significant improvements in public health. The apparent contradiction between vaccines with high efficacy, but low effectiveness, may reflect the difficulty in diagnosing influenza under clinical conditions and the large number of strains circulating in the population. In contrast, during an influenza pandemic, where a single strain of virus is responsible for illnesses, an effective vaccine could produce a large decrease in the number of cases and be highly effective in controlling an epidemic. However, such a vaccine would have to be produced and distributed rapidly to have maximum effect.

Effectiveness of Vaccine

The CDC reports that studies demonstrate that vaccination is a cost-effective counter-measure to seasonal outbreaks of influenza. However it is not perfect. A study led by Dr. David K. Shay in February, 2008 reported that:

> *"full immunisation against flu provided about a 75 percent effectiveness rate in preventing hospitalisations from influenza complications in the 2005-6 and 2006-7 influenza seasons."*

Influenza vaccine has been demonstrated to prevent disease and death, both in numerous controlled studies and in painstaking scientific reviews of these studies. However the rigor of the science of these studies has also been criticized. A 2006 Cochrane review of influenza vaccination in the elderly stated "The apparent high effectiveness of the vaccines in preventing death from all causes may reflect a baseline imbalance in health status and other systematic differences in the two groups of participants. In one observational study, a sharply lower risk of death or hospitalisation for pneumonia was seen in vaccinated persons:

Results: The relative risk of death for vaccinated persons compared with unvaccinated persons was 0.39 [95% confidence interval (95% CI), 0.33-0.47] before influenza season, 0.56 (0.52-0.61) during influenza season, and 0.74 (0.67-0.80) after influenza season. The relative risk of pneumonia hospitalisation was 0.72 (0.59-0.89) before, 0.82 (0.75-0.89) during, and 0.95 (0.85-1.07) after influenza season. Adjustment for diagnosis code variables resulted in estimates that were further from the null, in all time periods. A better vitality and lower pneumonia hospitalisation was thus observed in the vaccinated group in this non-randomised population study before, during, and after influenza season. This could, however, have been ascribed in whole or in part to self-selection bias, since vaccinated persons were already healthier before flu season. Some sub-populations have been assumed to benefit from vaccination in the absence of directly specific studies. For example, a 2008 Cochrane review of healthy children found "Influenza vaccines are efficacious in children older than two but little evidence is available for children under two.".

The CDC in 2010, after a review of extant studies, extended its guidelines to recommend that every child over 6 months be given the influenza vaccine. Vaccines were shown effective against the influenza strains they are designed to protect against, but this translated to only a modest impact on working days lost due to influenza-like infections in a 2007 Cochrane review on influenza vaccines in healthy adults. While a 2010 Cochrane review noted that "Influenza vaccines have a modest effect in reducing influenza symptoms and working days lost" it stated found no evidence of prevention of complications, such as pneumonia, or transmission.

The group most vulnerable to non-pandemic flu, the elderly, is also the least benefitted by the vaccine, with an average efficacy rate ranging from 40-50% at age 65, and only 15-30% past age 70. There are multiple reasons behind this steep decline in vaccine efficacy, the most common of which are the declining immunological function and frailty associated with advanced age. An influenza vaccine with four times the usual amount of antigen (Fluzone High Dose) has shown increased immune response in the elderly and has now been approved by the U.S. Food and Drug Administration (FDA). In a non-pandemic year, a person in the United States aged 50–64 is nearly ten times more likely to die an influenza-associated death than a younger person, and a person over age 65 is over ten times more likely to die an influenza-associated death than the 50–64 age group. Vaccination of those over age 65 reduces influenza-associated death by about 50%.

However, it is unlikely that the vaccine completely explains the results since elderly people who get vaccinated are probably more healthy and health-conscious than those who do not. Elderly participants randomised to a high-dose group (60 micrograms) had antibody levels 44 to 79 percent higher than did those who received the normal dose of vaccine. Elderly volunteers receiving the higher dose were more likely to achieve protective levels of antibody. As mortality is also high among infants who contract influenza, the household contacts and caregivers of infants should be vaccinated to reduce the risk of passing an influenza infection to the infant. Data from the years when Japan required annual flu vaccinations for school-aged children indicate that vaccinating children the group most likely to catch and spread the disease—has a strikingly positive effect on reducing mortality among older people: one life saved for every 420 children who received the flu vaccine. This may be due to herd immunity or to direct causes, such as individual older people not being exposed to influenza. For example, retired grandparents often risk infection by caring for their sick grandchildren in households where the parents can't take time off work or are sick themselves. In most years (16 of the 19 years before 2007), the flu vaccine strains have been a good match for the circulating strains. In other flu seasons like that of 2007/2008, the match was less useful. But even a mismatched vaccine can often provide some protection:

Antibodies made in response to vaccination with one strain of influenza viruses can provide protection against different, but related strains. A less than ideal match may result in reduced vaccine effectiveness against the variant viruses, but it still can provide enough protection to prevent or lessen illness severity and prevent flu-related complications.

In addition, it's important to remember that the influenza vaccine contains three virus strains so the vaccine can also protect against the other two viruses. For these reasons, even during seasons when there is a less than ideal match, CDC continues to recommend influenza vaccination. This is particularly important for people at high risk for serious flu complications and their close contacts.

Comparing Flu Shot to Nasal Spray

Flu vaccines are available either as:

TIV (flu shot (injection) of trivalent (three strains; usually A/H1N1, A/H3N2, and B) inactivated (killed) vaccine) or

LAIV (nasal spray (mist) of live attenuated influenza vaccine.)

TIV works by putting into the bloodstream those parts of three strains of flu virus that the body uses to create antibodies; while LAIV works by inoculating against those same three strains that have been genetically modified to minimise symptoms of illness. LAIV is not recommended for individuals under age 2 or over age 50, but might be comparatively more effective among children over age 2.

A military study on military personnel showed that flu shots yielded less illness than nasal spray. Based on one of the largest head-to-head studies comparing LAIV and TIV (which was conducted by the U.S. Armed Forces Surveillance Centre on military personnel who were stationed in the United States during three flu seasons from 2004 through 2007), investigators concluded that: "It may be prudent to use TIV in patients who were vaccinated at least once in the past 2 years but LAIV against pandemic strains may be more protective than inactivated vaccines, because the population will probably lack preexisting immunity."

High-dose Vaccine

A high-dose vaccine (Fluzone High-Dose) 4x the strength of standard flu vaccine was approved by the FDA in late 2009. This vaccine is intended for people 65 and over, who typically have weakened immune response due to normal aging. The vaccine produces a greater immune response than standard vaccine, but it is not yet known whether it provides greater protection against flu. Study results are expected in 2012. CDC recommends the high-dose vaccine for people 65 and over but expresses no preference between it and standard vaccine.

Vaccination Recommendations

Various public health organisations, including the World Health Organisation, have recommended that yearly influenza vaccination be routinely offered to patients at risk of complications of influenza and those individuals who live with or care for high-risk individuals, including:

- The elderly (UK recommendation is those aged 65 or above)
- Patients with chronic lung diseases (asthma, COPD, etc.)
- Patients with chronic heart diseases (congenital heart disease, chronic heart failure, ischaemic heart disease)
- Patients with chronic liver diseases (including cirrhosis)
- Patients with chronic renal diseases (such as the nephrotic syndrome)
- Patients who are immunosuppressed (those with HIV or who are receiving drugs to suppress the immune system such as chemotherapy and long-term steroids) and their household contacts

- People who live together in large numbers in an environment where influenza can spread rapidly, such as prisons, nursing homes, schools, and dormitories
- Healthcare workers (both to prevent sickness and to prevent spread to patients)
- *Pregnant women:* However, a 2009 review concluded that there was insufficient evidence to recommend routine use of trivalent influenza vaccine during the first trimester of pregnancy.
- Children from ages six months to two years.

Both types of flu vaccines are contraindicated for those with severe allergies to egg proteins and people with a history of Guillain-Barré syndrome.

Public Health Law Research, an independent organisation, published in 2009 several evidence briefs summarising the research assessing the effect of specific laws and policy on public health.

There is sufficient evidence supporting the effectiveness of requiring vaccinations as a condition for attending child care facilities and schools. There is insufficient evidence to assess the effectiveness of requiring vaccinations as a condition for specified jobs as a means of reducing incidence of specific diseases among particularly vulnerable populations.

There is strong evidence supporting the effectiveness of standing orders which allow healthcare workers without prescription authority to administer vaccines under defined circumstances as a public health intervention aimed at increasing vaccination rates.

Side Effects

- Side effects of the inactivated/dead flu vaccine injection include:
- Mild soreness, redness, and swelling where the shot was given
- Fever
- Aches
- These problems usually begin soon after the injection, and last 1–2 days.
- Side effects of the activated/live/LAIV flu nasal spray vaccine:
- Some children and adolescents 2–17 years of age have reported:
- Runny nose, nasal congestion or cough
- Fever

- Headache and muscle aches
- Wheesing
- Abdominal pain or occasional vomiting or diarrhea
- Some adults 18–49 years of age have reported:
- Runny nose or nasal congestion
- Sore throat
- Cough, chills, tiredness/weakness
- Headache
- More severe, but very rare side effects include:
- Life-threatening allergic reaction.

Some injection-based flu vaccines intended for adults in the United States contain thiomersal (also known as thimerosal). Despite some controversy in the media, the World Health Organisation has concluded that there is no evidence of toxicity from thiomersal in vaccines and no reason on grounds of safety to change to more-expensive single-dose administration. Although Guillain-Barré syndrome had been feared as a complication of vaccination, the CDC states that most studies on modern influenza vaccines have seen no link with Guillain-Barré.

A review has concluded that the 2009 H1N1 ("swine flu") vaccine has a safety profile similar to that of seasonal vaccine. Although one review gives an incidence of about one case per million vaccinations, a large study in China, reported in the NEJM covering close to 100 million doses of vaccine against the 2009 H1N1 "swine" flu found only eleven cases of Guillain-Barre syndrome, (0.1%) total incidence in persons vaccinated, actually lower than the normal rate of the disease in China, and no other notable side effects; "The risk-benefit ratio, which is what vaccines and everything in medicine is about, is overwhelmingly in favour of vaccination."

Getting infected by influenza itself increases both the risk of death (up to 1 in 10,000) and increases the risk of developing Guillain-Barré syndrome to a much higher level than the highest level of suspected vaccine involvement (approx. 10 times higher by recent estimates).

Flu Vaccine Manufacturing

Flu vaccine is usually grown in fertilised chicken eggs. In February preceding each fall's flu season (in the Northern hemisphere), three strains of flu are selected and chicken eggs inoculated.

As of November 2007, both the conventional injection and the nasal spray are manufactured using chicken eggs. The European Union has also approved Optaflu, a vaccine produced by Novartis using vats of animal cells. This technique is expected to be more scalable and avoid problems with eggs, such as allergic reactions and incompatibility with strains that affect avians like chickens.

A DNA-based vaccination, which is hoped to be even faster to manufacture, is currently in clinical trials, but has not yet been proven safe and effective. Research continues into the idea of a "universal" influenza vaccine (but no vaccine candidates have been announced) which would not need to be tailored to work on particular strains, but would be effective against a broad variety of influenza viruses.

In a 2007 report, the current global capacity of approximately 826 million seasonal influenza vaccine doses (inactivated and live) was double the current production of 413 million doses. In an aggressive scenario of producing pandemic influenza vaccines by 2013, only 2.8 billion courses could be produced in a six-month time frame. If all high- and upper-middle-income countries sought vaccines for their entire populations in a pandemic, nearly 2 billion courses would be required. If China pursued this goal as well, more than 3 billion courses would be required to serve these populations. Vaccine research and development is ongoing to identify novel vaccine approaches that could produce much greater quantities of vaccine at a price that is affordable to the global population.

An effective method of vaccine generation that bypasses the need for eggs is the construction of "influenza virus-like particle (VLP)". VLP is a non-egg, non-mammalian cell culture-based vaccine, purified from the supernatants of Spodoptera frugiperda Sf9 insect cells following infection of baculovirus vectors encoding an expression cassette made up of only three influenza virus structural proteins, hemagglutinin (HA), neuraminidase (NA), and matrix (M1) VLPs elicit antibodies that recognise a broader panel of antigenically distinct viral isolates compared to other vaccines in the hemagglutination-inhibition (HAI) assay.

H5N1

Vaccines have been formulated against several of the avian H5N1 influenza varieties. Vaccination of poultry against the ongoing H5N1 epizootic is widespread in certain countries. Some vaccines also exist for use in humans, and others are in testing, but none have been made available to civilian populations, nor produced in quantities sufficient to protect more than a tiny fraction of the Earth's population in the event of an H5N1 pandemic.

- Three H5N1 vaccines for humans have been licensed as of June 2008:
- Sanofi Pasteur's vaccine approved by the United States in April 2007,
- GlaxoSmithKline's vaccine Pandemrix approved by the European Union in May 2008, and
- CSL Limited's vaccine approved by Australia in June 2008.
- All are produced in eggs and would require many months to be altered to a pandemic version.

H5N1 continually mutates, meaning vaccines based on current samples of avian H5N1 cannot be depended upon to work in the case of a future pandemic of H5N1. While there can be some cross-protection against related flu strains, the best protection would be from a vaccine specifically produced for any future pandemic flu virus strain. Dr. Daniel Lucey, co-director of the Biohazardous Threats and Emerging Diseases graduate program at Georgetown University, has made this point, "There is no H5N1 pandemic so there can be no pandemic vaccine." However, "pre-pandemic vaccines" have been created; are being refined and tested; and do have some promise both in furthering research and preparedness for the next pandemic. Vaccine manufacturing companies are being encouraged to increase capacity so that if a pandemic vaccine is needed, facilities will be available for rapid production of large amounts of a vaccine specific to a new pandemic strain.

Problems with H5N1 vaccine production include:

- Lack of overall production capacity
- Lack of surge production capacity (it is impractical to develop a system that depends on hundreds of millions of 11 day old specialised eggs on a standby basis)
- The pandemic H5N1 might be lethal to chickens.

Cell culture (cell-based) manufacturing technology can be applied to influenza vaccines as they are with most viral vaccines and thereby solve the problems associated with creating flu vaccines using chicken eggs as is currently done.:

Currently, influenza vaccine for the annual, seasonal influenza program comes from four manufacturers. However, only a single manufacturer produces the annual vaccine entirely within the U.S. Thus, if a pandemic occurred and existing U.S.-based influenza vaccine manufacturing capacity was completely diverted to producing a pandemic vaccine, supply would be severely limited. Moreover, because the annual influenza manufacturing process takes place during most of the year, the time and capacity to produce vaccine against potential pandemic viruses for a stockpile, while continuing annual influenza vaccine production, is limited. Since supply will be limited, it is critical for HHS to be able to direct vaccine distribution in accordance with predefined groups; HHS will ensure the building of capacity and will engage states in a discussion about the purchase and distribution of pandemic influenza vaccine.

Vaccine production capacity: The protective immune response generated by current influenza vaccines is largely based on viral hemagglutinin (HA) and neuraminidase (NA) antigens in the vaccine. As a consequence, the basis of influenza vaccine manufacturing is growing massive quantities of virus in order to have sufficient amounts of these protein antigens to stimulate immune responses. Influenza vaccines used in the United States and around world are manufactured by growing virus in fertilised hens' eggs, a commercial process that has been in place for decades. To achieve current vaccine production targets millions of 11-day old fertilised eggs must be available every day of production.

In the near term, further expansion of these systems will provide additional capacity for the U.S.-based production of both seasonal and pandemic vaccines, however, the surge capacity that will be needed for a pandemic response cannot be met by egg-based vaccine production alone, as it is impractical to develop a system that depends on hundreds of millions of 11-day old specialised eggs on a standby basis. In addition, because a pandemic could result from an avian influenza strain that is lethal to chickens, it is impossible to ensure that eggs will be available to produce vaccine when needed.

In contrast, cell culture manufacturing technology can be applied to influenza vaccines as they are with most viral vaccines (e.g., polio vaccine, measles-mumps-rubella vaccine, chickenpox vaccine). In this system, viruses are grown in closed systems such as bioreactors containing large numbers of cells in growth media rather than eggs. The surge capacity afforded by cell-based technology is insensitive to seasons and can be adjusted to vaccine demand, as capacity can be increased or decreased by the number of bioreactors or the volume used within a bioreactor. In addition to supporting basic research on cell-based influenza vaccine development, HHS is currently supporting a number of vaccine manufacturers in the advanced development of cell-based influenza vaccines with the goal of developing U.S.-licensed cell-based influenza vaccines produced in the United States. The US government has purchased from Sanofi Pasteur and Chiron Corporation several million doses of vaccine meant to be used in case of an influenza pandemic of H5N1 avian influenza and is conducting clinical trials with these vaccines. Researchers at the University of Pittsburgh have had success with a genetically engineered vaccine that took only a month to make and completely protected chickens from the highly pathogenic H5N1 virus.

According to the United States Department of Health & Human Services

In addition to supporting basic research on cell-based influenza vaccine development, HHS is currently supporting a number of vaccine manufacturers in the advanced development of cell-based influenza vaccines with the goal of developing U.S.-licensed cell-based influenza vaccines produced in the United States. Dose-sparing technologies. Current U.S.-licensed vaccines stimulate an immune response based on the quantity of HA (hemagglutinin) antigen included in the dose. Methods to stimulate a strong immune response using less HA antigen are being studied in H5N1 and H9N2 vaccine trials. These include changing the mode of delivery from intramuscular to intradermal and the addition of immune-enhancing adjuvant to the vaccine formulation. Additionally, HHS is soliciting contract proposals from manufacturers of vaccines, adjuvants, and medical devices for the development and licensure of influenza vaccines that will provide dose-sparing alternative strategies.

Chiron Corporation is now recertified and under contract with the National Institutes of Health to produce 8,000–10,000 investigational doses of Avian Flu (H5N1) vaccine. MedImmune and Aventis Pasteur are under similar contracts.

The United States government hopes to obtain enough vaccine in 2006 to treat 4 million people. However, it is unclear whether this vaccine would be effective against a hypothetical mutated strain that would be easily transmitted through human populations, and the shelflife of stockpiled doses has yet to be determined. The *New England Journal of Medicine* reported on March 30, 2006 on one of dozens of vaccine studies currently being conducted. The Treanor et al. study was on vaccine produced from the human isolate (A/Vietnam/1203/2004 H5N1) of a virulent clade 1 influenza A (H5N1) virus with the use of a plasmid rescue system, with only the hemagglutinin and neuraminidase genes expressed and administered without adjuvant.

"The rest of the genes were derived from an avirulent egg-adapted influenza A/PR/8/34 strain.

The hemagglutinin gene was further modified to replace six basic amino acids associated with high pathogenicity in birds at the cleavage site between hemagglutinin 1 and hemagglutinin 2. Immunogenicity was assessed by microneu-tralisation and hemagglutination-inhibition assays with the use of the vaccine virus, although a subgroup of samples were tested with the use of the wild-type influenza A/Vietnam/1203/2004 (H5N1) virus."

The results of this study combined with others scheduled to be completed by spring 2007 is hoped will provide a highly immunogenic vaccine that is cross-protective against heterologous influenza strains.

On August 18, 2006. the World Health Organisation changed the H5N1 strains recommended for candidate vaccines for the first time since 2004. "The WHO's new prototype strains, prepared by reverse genetics, include three new H5N1 subclades.

The hemagglutinin sequences of most of the H5N1 avian influenza viruses circulating in the past few years fall into two genetic groups, or clades. Clade 1 includes human and bird isolates from Vietnam, Thailand, and Cambodia and bird isolates from Laos and Malaysia. Clade 2 viruses were first identified in bird isolates from China, Indonesia, Japan, and South Korea before spreading westward to the Middle East, Europe, and Africa.

The clade 2 viruses have been primarily responsible for human H5N1 infections that have occurred during late 2005 and 2006, according to WHO. Genetic analysis has identified six subclades of clade 2, three of which have a distinct geographic distribution and have been implicated in human infections:

- Subclade 1, Indonesia
- Subclade 2, Middle East, Europe, and Africa
- Subclade 3, China

On the basis of the three subclades, the WHO is offering companies and other groups that are interested in pandemic vaccine development these three new prototype strains:

- An A/Indonesia/2/2005-like virus
- An A/Bar headed goose/Quinghai/1A/2005-like virus
- An A/Anhui/1/2005-like virus.

Until now, researchers have been working on prepandemic vaccines for H5N1 viruses in clade 1. In March, the first clinical trial of a U.S. vaccine for H5N1 showed modest results. In May, French researchers showed somewhat better results in a clinical trial of an H5N1 vaccine that included an adjuvant. Vaccine experts aren't sure if a vaccine effective against known H5N1 viral strains would be effective against future strains. Although the new viruses will now be available for vaccine research, WHO said clinical trials using the clade 1 viruses should continue as an essential step in pandemic preparedness, because the trials yield useful information on priming, cross-reactivity, and cross-protection by vaccine viruses from different clades and subclades."

As of November 2006, the United States Department of Health and Human Services still had enough H5N1 pre-pandemic vaccine to treat about 3 million people (5.9 million full-potency doses) in spite of 0.2 million doses used for research and 1.4 million doses that have begun to lose potency (from the original 7.5 million full-potency doses purchased from Sanofi Pasteur and Chiron Corp.).

The expected shelf life of seasonal flu vaccine is about a year so the fact that most of the H5N1 pre-pandemic stockpile is still good after about 2 years is considered encouraging.

Annual Reformulation of Flu Vaccine

Each year, three strains are chosen for selection in that year's flu vaccination by the WHO Global Influenza Surveillance Network. The chosen strains are the H1N1, H3N2, and Type-B strains thought most likely to cause significant human suffering in the coming season.

Due to the high mutation rate of the virus a particular vaccine formulation is effective for at most about a year. The World Health Organisation coordinates the contents of the vaccine each year to contain the most likely strains of the virus to attack the next year.

"The WHO Global Influenza Surveillance Network was established in 1952. The network comprises 4 WHO Collaborating Centres (WHO CCs) and 112 institutions in 83 countries, which are recognised by

WHO as WHO National Influenza Centres (NICs). These NICs collect specimens in their country, perform primary virus isolation and preliminary antigenic characterisation.

They ship newly isolated strains to WHO CCs for high level antigenic and genetic analysis, the result of which forms the basis for WHO recommendations on the composition of influenza vaccine for the Northern and Southern Hemisphere each year."

The Global Influenza Surveillance Network's selection of viruses for the vaccine manufacturing process is based on its best estimate of which strains will be predominant the next year, amounting in the end to well-informed but fallible guesswork.

Formal WHO recommendations first issued in 1973; beginning 1999 there have been two recommendations per year, one for the northern hemisphere (N) and the other for the southern hemisphere (S).

History of past WHO seasonal influenza vaccine composition recommendations:

2002–2003 Northern Hemisphere Winter Season

The vaccines produced for the 2002–2003 season use:

- An A/New Caledonia/20/1999-like(H1N1);
- An A/Moscow/10/1999-like(H3N2);
- A B/Hong Kong/330/2001-like viruses.

2003 Southern Hemisphere Winter Season

The composition of influenza virus vaccines for use in the 2003 Southern Hemisphere influenza season recommended by the World Health Organisation was:

- An A/New Caledonia/20/99(H1N1)-like virus
- An A/Moscow/10/99(H3N2)-like virus (The widely used vaccine strain is A/Panama/2007/99)
- A B/Hong Kong/330/2001-like virus (Some vaccine strains used were B/Shandong/7/97, B/Hong Kong/330/2001, B/Hong Kong/1434/2002).

2003–2004 Northern Hemisphere Winter Season

The production of flu vaccine requires a lead time of about six months before the season. It is possible that by flu season a strain becomes common for which the vaccine does not provide protection. In the 2003–2004 season the vaccine was produced to protect against

A/Panama, A/New Caledonia, and B/Hong Kong. A new strain, A/Fujian, was discovered after production of the vaccine started and vaccination gave only partial protection against this strain.

Nature magazine reported that the Influenza Genome Sequencing Project, using phylogenetic analysis of 156 H3N2 genomes, "explains the appearance, during the 2003–2004 season, of the 'Fujian/411/2002'-like strain, for which the existing vaccine had limited effectiveness" as due to an epidemiologically significant reassortment.

"Through a reassortment event, a minor clade provided the haemagglutinin gene that later became part of the dominant strain after the 2002–2003 season. Two of our samples, A/New York/269/2003 (H3N2) and A/New York/32/2003 (H3N2), show that this minor clade continued to circulate in the 2003–2004 season, when most other isolates were reassortants."

According to the CDC

During the 2003–2004 influenza season, influenza A (H1), A (H3N2), and B viruses co-circulated worldwide, and influenza A (H3N2) viruses predominated.

Several Asian countries reported widespread outbreaks of avian influenza A (H5N1) among poultry. In Vietnam and Thailand, these outbreaks were associated with severe illnesses and deaths among humans.

In the United States, the 2003–2004 influenza season began earlier than most seasons, peaked in December, was moderately severe in terms of its impact on mortality, and was associated predominantly with influenza A (H3N2) viruses.

During September 28, 2003 – May 22, 2004, WHO and NREVSS collaborating laboratories in the United States tested 130,577 respiratory specimens for influenza viruses; 24,649 (18.9%) were positive. Of these, 24,393 (99.0%) were influenza A viruses, and 249 (1.0%) were influenza B viruses.

Among the influenza A viruses, 7,191 (29.5%) were subtyped; 7,189 (99.9%) were influenza A (H3N2) viruses, and two (0.1%) were influenza A (H1) viruses. The proportion of specimens testing positive for influenza first increased to >10% during the week ending October 25, 2003 (week 43), peaked at 35.2% during the week ending November 29 (week 48), and declined to <10% during the week ending January 17, 2004 (week 2).

The peak percentage of specimens testing positive for influenza during the previous four seasons had ranged from 23% to 31% and peaked during late December to late February.

As of June 15, 2004, CDC had antigenically characterised 1,024 influenza viruses collected by U.S. laboratories since October 1, 2003: 949 influenza A (H3N2) viruses, three influenza A (H1) viruses, one influenza A (H7N2) virus, and 71 influenza B viruses. Of the 949 influenza A (H3N2) isolates characterised, 106 (11.2%) were similar antigenically to the vaccine strain A/Panama/2007/99 (H3N2), and 843 (88.8%) were similar to the drift variant, A/Fujian/411/2002 (H3N2). Of the three A (H1) isolates that were characterised, two were H1N1 viruses, and one was an H1N2 virus. The hemagglutinin proteins of the influenza A (H1) viruses were similar antigenically to the hemagglutinin of the vaccine strain A/New Caledonia/20/99. Of the 71 influenza B isolates that were characterised, 66 (93%) belonged to the B/Yamagata/16/88 lineage and were similar antigenically to B/ Sichuan/379/99, and five (7%) belonged to the B/Victoria/2/87 lineage and were similar antigenically to the corresponding vaccine strain B/ Hong Kong/330/2001.

H9N2

In December 2003, one confirmed case of avian influenza A (H9N2) virus infection was reported in a child aged five years in Hong Kong.

The child had fever, cough, and nasal discharge in late November, was hospitalised for two days, and fully recovered. The source of this child's H9N2 infection is unknown.

H5N1

During January–March 2004, a total of 34 confirmed human cases of avian influenza A (H5N1) virus infection were reported in Vietnam and Thailand. The cases were associated with severe respiratory illness requiring hospitalisation and a case-fatality proportion of 68% (Vietnam: 22 cases, 15 deaths; Thailand: 12 cases, eight deaths). A substantial proportion of the cases were among children and young adults (i.e., persons aged 5–24 years). These cases were associated with widespread outbreaks of highly pathogenic H5N1 influenza among domestic poultry.

H7N3

During March 2004, health authorities in Canada reported two confirmed cases of avian influenza A (H7N3) virus infection in poultry workers who were involved in culling of poultry during outbreaks of highly pathogenic H7N3 on farms in the Fraser River Valley, British Columbia. One patient had unilateral conjunctivitis and nasal discharge, and the other had unilateral conjunctivitis and headache. Both illnesses resolved without hospitalisation.

H7N2

During the 2003–2004 influenza season, a case of avian influenza A (H7N2) virus infection was detected in an adult male from New York, who was hospitalised for upper and lower respiratory tract illness in November 2003.

Influenza A (H7N2) virus was isolated from a respiratory specimen from the patient, whose acute symptoms resolved.

The source of this person's infection is unknown.

2004 Southern Hemisphere Winter Season

The composition of influenza virus vaccines for use in the 2004 Southern Hemisphere influenza season recommended by the World Health Organisation was:

- An A/New Caledonia/20/99(H1N1)-like virus
- An A/Fujian/411/2002(H3N2)-like virus (A/Kumamoto/102/2002 and A/Wyoming/3/2003 were egg-grown A/Fujian/411/2002-like viruses)
- a B/Hong Kong/330/2001-like virus (B/Shandong/7/97, B/Hong Kong/330/2001 and B/Hong Kong/1434/2002 were among those used at the time. B/Brisbane/32/2002 was also available.)

2004–2005 Northern Hemisphere Winter Season

On the basis of antigenic analyses of recently isolated influenza viruses, epidemiologic data, and postvaccination serologic studies in humans, the Food and Drug Administration's Vaccines and Related Biological Products Advisory Committee (VRBPAC) recommended that the 2004–05 trivalent influenza vaccine for the United States contain A/New Caledonia/20/99-like (H1N1), A/ Fujian/411/2002-like (H3N2), and B/Shanghai/361/2002-like viruses.

Because of the growth properties of the A/Wyoming/3/2003 and B/ Jiangsu/10/2003 viruses, U.S. vaccine manu-facturers are using these antigenically equivalent strains in the vaccine as the H3N2 and B components, respectively. The A/New Caledonia/20/99 virus will be retained as the H1N1 component of the vaccine.

2005 Southern Hemisphere Winter Season

The composition of influenza virus vaccines for use in the 2005 Southern Hemisphere influenza season recommended by the World Health Organisation was:

- An A/New Caledonia/20/99(H1N1)-like virus;
- An A/Wellington/1/2004(H3N2)-like virus;
- A B/Shanghai/361/2002-like virus (B/Shanghai/361/2002, B/Jilin/20/2003 and B/Jiangsu/10/2003 were used at the time)

2005–2006 Northern Hemisphere Winter Season

The vaccines produced for the 2005–2006 season use:

- An A/New Caledonia/20/1999-like(H1N1);
- An A/California/7/2004-like(H3N2) (or the antigenically equivalent strain A/New York/55/2004);
- A B/Jiangsu/10/2003-like viruses.

In people in the United States, overall flu and pneumonia deaths were below those of a typical flu season with 84% Influenzavirus A and the rest Influenzavirus B. Of the patients who had Type A viruses, 80% had viruses identical or similar to the A bugs in the vaccine. 70% of the people testing positive for a B virus had Type B Victoria, a version not found in the vaccine. "During the 2005–06 season, influenza A (H3N2) viruses predo-minated overall, but late in the season influenza B viruses were more frequently isolated than influenza A viruses. Influenza A (H1N1) viruses circulated at low levels throughout the season. Nationally, activity was low from October through early January, increased during February, and peaked in early March. Peak activity was less intense, but activity remained elevated for a longer period of time this season compared to the previous three seasons. The longer period of elevated activity may be due in part to regional differences in the timing of peak activity and intensity of influenza B activity later in the season."

2006 Southern Hemisphere Winter Season

The composition of influenza virus vaccines for use in the 2006 Southern Hemisphere influenza season recommended by the World Health Organisation was:

- An A/New Caledonia/20/99(H1N1)-like virus;
- An A/California/7/2004(H3N2)-like virus (A/New York/55/2004 was used at the time);
- A B/Malaysia/2506/2004-like virus.

2006–2007 Northern Hemisphere Winter Season

The 2006–2007 influenza vaccine composition recommended by the World Health Organisation on February 15, 2006 and the U.S. FDA's Vaccines and Related Biological Products Advisory Committee (VRBPAC) on February 17, 2006 use:

- An A/New Caledonia/20/99 (H1N1)-like virus;
- An A/Wisconsin/67/2005 (H3N2)-like virus (A/Wisconsin/67/2005 and A/Hiroshima/52/2005 strains);
- A B/Malaysia/2506/2004-like virus from B/Malaysia/2506/2004 and B/Ohio/1/2005 strains which are of B/Victoria/2/87 lineage.

2007 Southern Hemisphere Winter Season

The composition of influenza virus vaccines for use in the 2007 Southern Hemisphere influenza season recommended by the World Health Organisation on September 20, 2006 was:

- An A/New Caledonia/20/99(H1N1)-like virus,
- An A/Wisconsin/67/2005(H3N2)-like virus (A/Wisconsin/67/2005 and A/Hiroshima/52/2005 were used at the time),
- A B/Malaysia/2506/2004-like virus.

2007–2008 Northern Hemisphere Winter Season

The composition of influenza virus vaccines for use in the 2007–2008 Northern Hemisphere influenza season recommended by the World Health Organisation on February 14, 2007 was:

- An A/Solomon Islands/3/2006 (H1N1)-like virus;
- An A/Wisconsin/67/2005 (H3N2)-like virus (A/Wisconsin/67/2005 (H3N2) and A/Hiroshima/52/2005 were used at the time);
- A B/Malaysia/2506/2004-like virus.

In the US, the CDC reported in Feb 2008 that the H1N1 component was a good match (96%) to the infections occurring. But 87% of the H3N2 are A/Brisbane/10/2007-like viruses, which are a recent antigenic variant of the vaccine strain, A/Wisconsin. And 93% of the B viruses are in a B/Yamagata lineage that is relatively distinct from the vaccine strain B/Victoria lineage. Only one of the three components was a good match; A/Wisconsin is moderately protective against the drifted A/Brisbane strain. 4.5% of those viruses tested are resistant to Oseltamivir, or Tamiflu—a significant increase over previous years. This vaccine has been described as 40% effective compared to other years that have been 85–95% effective.

2008 Southern Hemisphere Winter Season

The composition of virus vaccines for use in the 2008 Southern Hemisphere influenza season recommended by the World Health Organisation on September 17–19, 2007 was:

- An A/Solomon Islands/3/2006 (H1N1)-like virus;
- An A/Brisbane/10/2007 (H3N2)-like virus;
- A B/Florida/4/2006-like virus

2008-2009 Northern Hemisphere Winter Season

The composition of virus vaccines for use in the 2008-2009 Northern Hemisphere influenza season recommended by the World Health Organisation on February 14, 2008 was:

- An A/Brisbane/59/2007 (H1N1)-like virus;
- An A/Brisbane/10/2007 (H3N2)-like virus;
- a B/Florida/4/2006-like virus (B/Florida/4/2006 and B/Brisbane/3/2007 (a B/Florida/4/2006-like virus) were used at the time).

As of May 30, 2009: "CDC has antigenically characterised 1,567 seasonal human influenza viruses [947 influenza A (H1), 162 influenza A (H3) and 458 influenza B viruses] collected by U.S. laboratories since October 1, 2008, and 84 novel influenza A (H1N1) viruses. All 947 influenza seasonal A (H1) viruses are related to the influenza A (H1N1) component of the 2008-09 influenza vaccine (A/Brisbane/59/2007). All 162 influenza A (H3N2) viruses are related to the A (H3N2) vaccine component (A/Brisbane/10/2007).

All 84 novel influenza A (H1N1) viruses are related to the A/California/07/2009 (H1N1) reference virus selected by WHO as a potential candidate for novel influenza A (H1N1) vaccine. Influenza B viruses currently circulating can be divided into two distinct lineages represented by the B/Yamagata/16/88 and B/Victoria/02/87 viruses.

Sixty-one influenza B viruses tested belong to the B/Yamagata lineage and are related to the vaccine strain (B/Florida/04/2006). The remaining 397 viruses belong to the B/Victoria lineage and are not related to the vaccine strain."

2009 Southern Hemisphere Winter Season

The composition of virus vaccines for use in the 2009 Southern Hemisphere influenza season recommended by the World Health Organisation on September 17–19, 2008 was:

- An A/Brisbane/59/2007 (H1N1)-like virus;
- An A/Brisbane/10/2007 (H3N2)-like virus;
- A B/Florida/4/2006-like virus.

2009-2010 Northern Hemisphere Winter Season

The composition of virus vaccines for use in the 2009-2010 Northern Hemisphere influenza season recommended by the World Health Organisation on February 12, 2009 was:

- An A/Brisbane/59/2007 (H1N1)-like virus;
- An A/Brisbane/10/2007 (H3N2)-like virus;
- A B/Brisbane/60/2008-like virus.

Since the A/Brisbane/59/2007 (H1N1)-like virus used in the vaccine is an unrelated seasonal strain of influenza, it probably cannot create immunity to the new, non-seasonal strain of influenza A virus subtype H1N1 responsible for the 2009 swine flu outbreak.

2010 Southern Hemisphere Winter Season

- An A/California/7/2009 (H1N1)-like virus;
- An A/Perth/16/2009 (H3N2)-like virus;
- A B/Brisbane/60/2008-like virus.

The H1N1 strain used in this composition is the same strain used in the 2009 flu pandemic vaccine.

2010-2011 Northern Hemisphere Winter Season

The composition of virus vaccines for use in the 2010-2011 Northern Hemisphere influenza season recommended by the World Health Organisation on February 18, 2010 was:

- An A/California/7/2009 (H1N1)-like virus;
- An A/Perth/16/2009 (H3N2)-like virus;
- A B/Brisbane/60/2008-like virus.

The H1N1 strain used in this composition is the same strain used in the 2009 flu pandemic vaccine.

2011 Southern Hemisphere Winter Season

The composition of virus vaccines for use in the 2011 Southern Hemisphere influenza season recommended by the World Health Organisation on September 29, 2010 was:

- An A/California/7/2009 (H1N1)-like virus;
- An A/Perth/16/2009 (H3N2)-like virus;
- A B/Brisbane/60/2008-like virus.

The H1N1 strain used in this composition is the same strain used in the 2009 flu pandemic vaccine.

2011-2012 Northern Hemisphere Winter Season

The composition of virus vaccines for use in the 2011-2012 Northern Hemisphere influenza season recommended by the World Health Organisation on February 17, 2011 was:

- An A/California/7/2009 (H1N1)-like virus;
- An A/Perth/16/2009 (H3N2)-like virus;
- A B/Brisbane/60/2008-like virus.

The H1N1 strain used in this composition is the same strain used in the 2009 flu pandemic vaccine.

Flu Vaccine for Nonhumans

"Vaccination in the veterinary world pursues four goals: (i) protection from clinical disease, (ii) protection from infection with virulent virus, (iii) protection from virus excretion, and (iv) serological differentiation of infected from vaccinated animals (so-called DIVA principle). In the field of influenza vaccination, neither commercially available nor experimentally tested vaccines have been shown so far to fulfil all of these requirements."

Horses

Horses with horse flu can run a fever, have a dry hacking cough, have a runny nose, and become depressed and reluctant to eat or drink for several days but usually recover in two to three weeks. "Vaccination schedules generally require a primary course of 2 doses, 3–6 weeks apart, followed by boosters at 6–12 month intervals. It is generally recognised that in many cases such schedules may not maintain protective levels of antibody and more frequent administration is advised in high-risk situations."

It is a common requirement at shows in the United Kingdom that horses be vaccinated against equine flu and a vaccination card must be produced; the FEI requires vaccination every six months.

Poultry

Poultry vaccines for bird flu are made on the cheap and are not filtered and purified like human vaccines to remove bits of bacteria or other viruses. They usually contain whole virus, not just hemagglutinin as in most human flu vaccines. Purification to standards needed for humans is far more expensive than the original creation of the unpurified vaccine from eggs. There is no market for veterinary vaccines that are that expensive. Another difference between human and poultry vaccines is that poultry vaccines are adjuvated with mineral oil, which induces a strong immune reaction but can cause inflammation and abscesses. "Chicken vaccinators who have accidentally jabbed themselves have developed painful swollen fingers or even lost thumbs, doctors said. Effectiveness may also be limited. Chicken vaccines are often only vaguely similar to circulating flu strains — some contain an H5N2 strain isolated in Mexico years ago. 'With a chicken, if you use a vaccine that's only 85 percent related, you'll get protection,' Dr. Cardona said. 'In humans, you can get a single point mutation, and a vaccine that's 99.99 percent related won't protect you.' And they are weaker than human vaccines. 'Chickens are smaller and you only need to protect them for six weeks, because that's how long they live till you eat them,' said Dr. John J.

Treanor, a vaccine expert at the University of Rochester. Human seasonal flu vaccines contain about 45 micrograms of antigen, while an experimental A(H5N1) vaccine contains 180. Chicken vaccines may contain less than 1 microgram. 'You have to be careful about extrapolating data from poultry to humans,' warned Dr. David E. Swayne, director of the agriculture department's Southeast Poultry Research Laboratory. 'Birds are more closely related to dinosaurs.'"

Researchers, led by Nicholas Savill of the University of Edinburgh in Scotland, used mathematical models to simulate the spread of H5N1 and concluded that "at least 95 per cent of birds need to be protected to prevent the virus spreading silently. In practice, it is difficult to protect more than 90 per cent of a flock; protection levels achieved by a vaccine are usually much lower than this." The Food and Agriculture Organisation of the United Nations has issued recommendations on the prevention and control of avian influenza in poultry, including the use of vaccination. A filtered and purified Influenza A vaccine for humans is being developed and many countries have recom-mended it be stockpiled so if an Avian influenza pandemic starts jumping to humans, the vaccine can quickly be administered to avoid loss of life. Avian influenza is sometimes called avian flu, and commonly bird flu.

Pigs

Swine origin influenza virus (SoIV) vaccines are extensively used in the swine industry in Europe and North America. Most swine flu vaccine manufacturers include an H1N1 and an H3N2 SoIV strains. Swine influenza has become a greater problem in recent decades. Evolution of the virus has resulted in inconsistent responses to traditional vaccines. Standard commercial swine origin flu vaccines are effective in controlling the problem when the virus strains match enough to have significant cross-protection and custom (autogenous) vaccines made from the specific viruses isolated are created and used in the more difficult cases. SoIV vaccine manufacture Novartis paints this picture: "A strain of swine origin influenza virus (SoIV) called H3N2, first identified in the US in 1998, has brought exasperating production losses to swine producers. Abortion storms are a common sign. Sows go off feed for two or three days and run a fever up to 106°F. Mortality in a naïve herd can run as high as 15%."

Dogs

In 2004, Influenza A virus subtype H3N8 was discovered to cause canine influenza. Because of the lack of previous exposure to this virus, dogs have no natural immunity to this virus. However a vaccine is now available.

Computer-assisted Vaccine Design

A new parametre has been defined to quantify the antigenic distance between two H3N2 influenza strains. This parametre was used to measure antigenic distance between circulating H3N2 strains and the closest vaccine component of the influenza vaccine. For the data between 1971 and 2004, the measure of antigenic distance correlated better with efficacy in humans of the H3N2 influenza A annual vaccine than did current measures of antigenic distance such as phylogenetic sequence analysis or ferret antisera inhibition assays. This measure of antigenic distance could be used to guide the design of the annual flu vaccine. The antigenic distance combined with a multiple-strain avian influenza transmission model was used to study the threat of simultaneous introduction of multiple avian influenza strains. Population at Risk (PaR) can be used to quantify the risk of a flu pandemic and to calculate the improvement that a multiple vaccine offers.

Influenza A Virus Subtype H5N1

Influenza A virus subtype H5N1, also known as "bird flu", A(H5N1) or simply H5N1, is a subtype of the influenza A virus which can cause illness in humans and many other animal species. A bird-adapted strain of H5N1, called HPAI A(H5N1) for "highly pathogenic avian influenza virus of type A of subtype H5N1", is the causative agent of H5N1 flu, commonly known as "avian influenza" or "bird flu". It is enzootic in many bird populations, especially in Southeast Asia. One strain of HPAI A(H5N1) is spreading globally after first appearing in Asia. It is epizootic (an epidemic in nonhumans) and panzootic (affecting animals of many species, especially over a wide area), killing tens of millions of birds and spurring the culling of hundreds of millions of others to stem its spread. Most references to "bird flu" and H5N1 in the popular media refer to this strain.

According to the FAO Avian Influenza Disease Emergency Situation Update, H5N1 pathogenicity is continuing to gradually rise in endemic areas but the avian influenza disease situation in farmed birds is being held in check by vaccination.

Eleven outbreaks of H5N1 were reported worldwide in June 2008 in five countries (China, Egypt, Indonesia, Pakistan and Vietnam) compared to 65 outbreaks in June 2006 and 55 in June 2007. The "global HPAI situation can be said to have improved markedly in the first half of 2008 but cases of HPAI are still underestimated and underreported in many countries because of limitations in country disease surveillance systems". On December 9, 2010 the WHO announced a total of 510 human cases which resulted in the deaths of 303 people since 2003.

A filtered and purified Influenza A vaccine for humans is being developed and many countries have recommended it be stockpiled so, if an Avian influenza pandemic starts jumping to humans, the vaccine can quickly be administered to avoid loss of life. Avian influenza is sometimes called avian flu, and commonly bird flu.

Overview

HPAI A(H5N1) is considered an avian disease, although there is some evidence of limited human-to-human transmission of the virus. A risk factor for contracting the virus is handling of infected poultry, but transmission of the virus from infected birds to humans is inefficient. Still, around 60% of humans known to have been infected with the current Asian strain of HPAI A(H5N1) have died from it, and H5N1 may mutate or reassort into a strain capable of efficient human-to-human transmission. In 2003, world-renowned virologist Robert G. Webster published an article titled "The world is teetering on the edge of a pandemic that could kill a large fraction of the human population" in *American Scientist*. He called for adequate resources to fight what he sees as a major world threat to possibly billions of lives. On September 29, 2005, David Nabarro, the newly appointed Senior United Nations System Coordinator for Avian and Human Influenza, warned the world that an outbreak of avian influenza could kill anywhere between 5 million and 150 million people. Experts have identified key events (creating new clades, infecting new species, spreading to new areas) marking the progression of an avian flu virus towards becoming pandemic, and many of those key events have occurred more rapidly than expected.

Due to the high lethality and virulence of HPAI A(H5N1), its endemic presence, its increasingly large host reservoir, and its significant ongoing mutations, the H5N1 virus is the world's largest current pandemic threat and billions of dollars are being spent researching H5N1 and preparing for a potential influenza pandemic. At least 12 companies and 17 governments are developing pre-pandemic influenza vaccines in 28 different clinical trials that, if successful, could turn a deadly pandemic infection into a nondeadly one. Full-scale production of a vaccine that could prevent any illness at all from the strain would require at least three months after the virus's emergence to begin, but it is hoped that vaccine production could increase until one billion doses were produced by one year after the initial identification of the virus.

H5N1 may cause more than one influenza pandemic as it is expected to continue mutating in birds regardless of whether humans develop herd immunity to a future pandemic strain. Influenza pandemics from its genetic offspring may include influenza A virus subtypes other than H5N1. While genetic analysis of the H5N1 virus shows that influenza pandemics from its genetic offspring can easily be far more lethal than the Spanish flu pandemic, planning for a future influenza pandemic is based on what can be done and there is no higher Pandemic Severity Index level than a Category 5 pandemic which, roughly speaking, is any pandemic as bad as the Spanish flu or worse; and for which *all* intervention measures are to be used.

Signs and Symptoms

The avian influenza hemagglutinin binds alpha 2-3 sialic acid receptors, while human influenza hemagglutinins bind alpha 2-6 sialic acid receptors. This means when the H5N1 strain infects humans, it will replicate in the lower respiratory tract, and consequently will cause viral pneumonia. There is as yet no human form of H5N1, so all humans who have caught it so far have caught avian H5N1. In general, humans who catch a humanised influenza A virus (a human flu virus of type A) usually have symptoms that include fever, cough, sore throat, muscle aches, conjunctivitis, and, in severe cases, breathing problems and pneumonia that may be fatal. The severity of the infection depends in large part on the state of the infected person's immune system and whether the victim has been exposed to the strain before (in which case they would be partially immune). No one knows if these or other symptoms will be the symptoms of a humanised H5N1 flu.

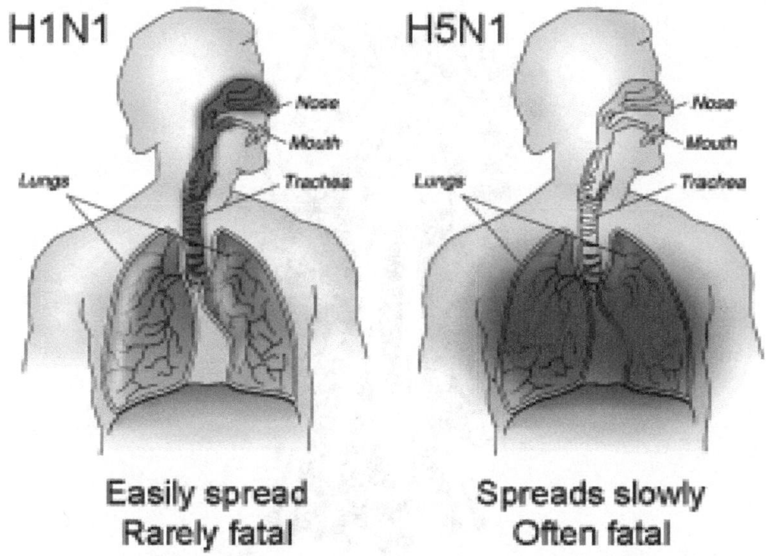

Figure: *The different sites of infection of seasonal H1N1 versus avian H5N1. This influences their lethality and ability to spread.*

The reported mortality rate of highly pathogenic H5N1 avian influenza in a human is high; WHO data indicate 60% of cases classified as H5N1 resulted in death. However, there is some evidence the actual mortality rate of avian flu could be much lower, as there may be many people with milder symptoms who do not seek treatment and are not counted.

In one case, a boy with H5N1 experienced diarrhea followed rapidly by a coma without developing respiratory or flu-like symptoms. There have been studies of the levels of cytokines in humans infected by the H5N1 flu virus. Of particular concern is elevated levels of tumour necrosis factor-alpha, a protein associated with tissue destruction at sites of infection and increased production of other cytokines. Flu virus-induced increases in the level of cytokines is also associated with flu symptoms, including fever, chills, vomiting and headache. Tissue damage associated with pathogenic flu virus infection can ultimately result in death. The inflammatory cascade triggered by H5N1 has been called a 'cytokine storm' by some, because of what seems to be a positive feedback process of damage to the body resulting from immune system stimulation. H5N1 induces higher levels of cytokines than the more common flu virus types.

Genetics

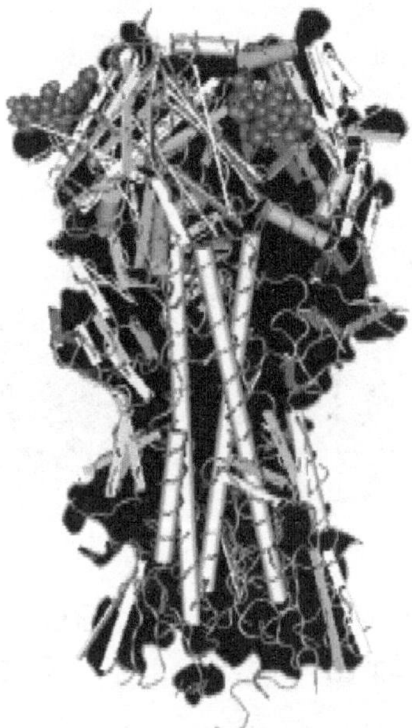

Figure : *The H in H5N1 stands for "Hemagglutinin", as depicted in this molecular model*

The first known strain of HPAI A(H5N1) (called A/chicken/ Scotland/59) killed two flocks of chickens in Scotland in 1959; but that strain was very different from the current highly pathogenic strain of H5N1. The dominant strain of HPAI A(H5N1) in 2004 evolved from 1999 to 2002 creating the Z genotype. It has also been called "Asian lineage HPAI A(H5N1)".

Asian lineage HPAI A(H5N1) is divided into two antigenic clades. "Clade 1 includes human and bird isolates from Vietnam, Thailand, and Cambodia and bird isolates from Laos and Malaysia. Clade 2 viruses were first identified in bird isolates from China, Indonesia,

Japan, and South Korea before spreading westward to the Middle East, Europe, and Africa. The clade 2 viruses have been primarily responsible for human H5N1 infections that have occurred during late 2005 and 2006, according to WHO. Genetic analysis has identified six subclades of clade 2, three of which have a distinct geographic distribution and have been implicated in human infections:

- Subclade 1, Indonesia
- Subclade 2, Europe, Middle East, and Africa (called EMA)
- Subclade 3, China"

A 2007 study focused on the EMA subclade has shed further light on the EMA mutations. "The 36 new isolates reported here greatly expand the amount of whole-genome sequence data available from recent avian influenza (H5N1) isolates.

Before our project, GenBank contained only 5 other complete genomes from Europe for the 2004–2006 period, and it contained no whole genomes from the Middle East or northern Africa. Our analysis showed several new findings.

First, all European, Middle Eastern, and African samples fall into a clade that is distinct from other contemporary Asian clades, all of which share common ancestry with the original 1997 Hong Kong strain. Phylogenetic trees built on each of the 8 segments show a consistent picture of 3 lineages.

Two of the clades contain exclusively Vietnamese isolates; the smaller of these, with 5 isolates, we label V1; the larger clade, with 9 isolates, is V2. The remaining 22 isolates all fall into a third, clearly distinct clade, labelled EMA, which comprises samples from Europe, the Middle East, and Africa. Trees for the other 7 segments display a similar topology, with clades V1, V2, and EMA clearly separated in each case. Analyses of all available complete influenza (H5N1) genomes and of 589 HA sequences placed the EMA clade as distinct from the major clades circulating in People's Republic of China, Indonesia, and Southeast Asia."

Terminology

- H5N1 isolates are identified like this actual HPAI A(H5N1) example, A/chicken/Nakorn-Patom/Thailand/CU-K2/04(H5N1):
- A stands for the species of influenza (A, B or C).
- chicken is the species the isolate was found in Nakorn-Patom/ Thailand is the place this specific virus was isolated

- CU-K2 identifies it from other influenza viruses isolated at the same place 04 represents the year 2004
- H5 stands for the fifth of several known types of the protein hemagglutinin.
- N1 stands for the first of several known types of the protein neuraminidase.

Other examples include: A/duck/Hong Kong/308/78(H5N3), A/avian/NY/01(H5N2), A/chicken/Mexico/31381-3/94(H5N2), and A/shoveler/Egypt/03(H5N2). As with other avian flu viruses, H5N1 has strains called "highly pathogenic" (HP) and "low-pathogenic" (LP). Avian influenza viruses that cause HPAI are highly virulent, and mortality rates in infected flocks often approach 100%. LPAI viruses have negligible virulence, but these viruses can serve as progenitors to HPAI viruses. The current strain of H5N1 responsible for the deaths of birds across the world is an HPAI strain; all other current strains of H5N1, including a North American strain that causes no disease at all in any species, are LPAI strains. All HPAI strains identified to date have involved H5 and H7 subtypes. The distinction concerns pathogenicity in poultry, not humans. Normally a highly pathogenic avian virus is not highly pathogenic to either humans or non-poultry birds. This current deadly strain of H5N1 is unusual in being deadly to so many species, including some, like domestic cats, never previously susceptible to any influenza virus.

Genetic Structure and Related Subtypes

H5N1 is a subtype of the species *Influenza A virus* of the *Influenzavirus A* genus of the *Orthomyxoviridae* family. Like all other influenza A subtypes, the H5N1 subtype is an RNA virus. It has a segmented genome of eight negative sense, single-strands of RNA, abbreviated as PB2, PB1, PA, HA, NP, NA, MP and NS.

HA codes for hemagglutinin, an antigenic glycoprotein found on the surface of the influenza viruses and is responsible for binding the virus to the cell that is being infected. NA codes for neuraminidase, an antigenic glycosylated enzyme found on the surface of the influenza viruses. It facilitates the release of progeny viruses from infected cells. The hemagglutinin (HA) and neuraminidase (NA) RNA strands specify the structure of proteins that are most medically relevant as targets for antiviral drugs and antibodies. HA and NA are also used as the basis for the naming of the different subtypes of influenza A viruses. This is where the *H* and *N* come from in *H5N1*.

Influenza A viruses are significant for their potential for disease and death in humans and other animals. Influenza A virus subtypes that have been confirmed in humans, in order of the number of known human pandemic deaths that they have caused, include:

- H1N1, which caused the 1918 flu pandemic ("Spanish flu") and currently is causing seasonal human flu and the 2009 flu pandemic ("swine flu")
- H2N2, which caused "Asian flu"
- H3N2, which caused "Hong Kong flu" and currently causes seasonal human flu

H5N1, ("bird flu"), which is noted for having a strain (Asian-linage HPAI H5N1) that kills over half the humans it infects, infecting and killing species that were never known to suffer from influenza viruses before (e.g. cats), being unable to be stopped by culling all involved poultry - some think due to being endemic in wild birds, and causing billions of dollars to be spent in flu pandemic preparation and preventiveness

- H7N7, which has unusual zoonotic potential and killed one person
- H1N2, which is currently endemic in humans and pigs and causes seasonal human flu
- H9N2, which has infected three people
- H7N2, which has infected two people
- H7N3, which has infected two people
- H10N7, which has infected two people.

Low Pathogenic H5N1

Low pathogenic avian influenza H5N1 (LPAI H5N1) also called "North American" H5N1 commonly occurs in wild birds. In most cases, it causes minor sickness or no noticeable signs of disease in birds. It is not known to affect humans at all.

The only concern about it is that it is possible for it to be transmitted to poultry and in poultry mutate into a highly pathogenic strain.

- 1975 – LPAI H5N1 was detected in a wild mallard duck and a wild blue goose in Wisconsin.
- 1981 and 1985 – LPAI H5N1 was detected in ducks by the University of Minnesota conducting a sampling procedure in which sentinel ducks were monitored in cages placed in the wild for a short period of time.

- 1983 – LPAI H5N1 was detected in ring-billed gulls in Pennsylvania.
- 1986 - LPAI H5N1 was detected in a wild mallard duck in Ohio.
- 2005 - LPAI H5N1 was detected in ducks in Manitoba, Canada.
- 2008 - LPAI H5N1 was detected in ducks in New Zealand.
- 2009 - LPAI H5N1 was detected in commercial poultry in British Columbia.

"In the past, there was no requirement for reporting or tracking LPAI H5 or H7 detections in wild birds so states and universities tested wild bird samples independently of USDA. Because of this, the above list of previous detections might not be all inclusive of past LPAI H5N1 detections.

However, the World Organisation for Animal Health (OIE) recently changed its requirement of reporting detections of avian influenza. Effective in 2006, all confirmed LPAI H5 and H7 AI subtypes must be reported to the OIE because of their potential to mutate into highly pathogenic strains. Therefore, USDA now tracks these detections in wild birds, backyard flocks, commercial flocks and live bird markets."

High Mutation Rate

Influenza viruses have a relatively high mutation rate that is characteristic of RNA viruses. The segmentation of its genome facilitates genetic recombination by segment reassortment in hosts infected with two different influenza viruses at the same time. A previously uncontagious strain may then be able to pass between humans, one of several possible paths to a pandemic.

The ability of various influenza strains to show species-selectivity is largely due to variation in the hemagglutinin genes. Genetic mutations in the hemagglutinin gene that cause single amino acid substitutions can significantly alter the ability of viral hemagglutinin proteins to bind to receptors on the surface of host cells.

Such mutations in avian H5N1 viruses can change virus strains from being inefficient at infecting human cells to being as efficient in causing human infections as more common human influenza virus types. This doesn't mean that one amino acid substitution can cause a pandemic, but it does mean that one amino acid substitution can cause an avian flu virus that is not pathogenic in humans to become pathogenic in humans.

Influenza A virus subtype H3N2 is endemic in pigs in China, and has been detected in pigs in Vietnam, increasing fears of the emergence of new variant strains. The dominant strain of annual flu virus in January 2006 was H3N2, which is now resistant to the standard antiviral drugs amantadine and rimantadine. The possibility of H5N1 and H3N2 exchanging genes through reassortment is a major concern. If a reassortment in H5N1 occurs, it might remain an H5N1 subtype, or it could shift subtypes, as H2N2 did when it evolved into the Hong Kong Flu strain of H3N2. Both the H2N2 and H3N2 pandemic strains contained avian influenza virus RNA segments. "While the pandemic human influenza viruses of 1957 (H2N2) and 1968 (H3N2) clearly arose through reassortment between human and avian viruses, the influenza virus causing the 'Spanish flu' in 1918 appears to be entirely derived from an avian source".

Prevention

There are several H5N1 vaccines for several of the avian H5N1 varieties, but the continual mutation of H5N1 renders them of limited use to date: while vaccines can sometimes provide cross-protection against related flu strains, the best protection would be from a vaccine specifically produced for any future pandemic flu virus strain. Dr. Daniel Lucey, co-director of the Biohazardous Threats and Emerging Diseases graduate program at Georgetown University has made this point, "There is no H5N1 pandemic so there can be no pandemic vaccine". However, "pre-pandemic vaccines" have been created; are being refined and tested; and do have some promise both in furthering research and preparedness for the next pandemic. Vaccine manufacturing companies are being encouraged to increase capacity so that if a pandemic vaccine is needed, facilities will be available for rapid production of large amounts of a vaccine specific to a new pandemic strain.

Public Health

"The United States is collaborating closely with eight international organisations, including the World Health Organisation (WHO), the Food and Agriculture Organisation of the United Nations (FAO), the World Organisation for Animal Health (OIE), and 88 foreign governments to address the situation through planning, greater monitoring, and full transparency in reporting and investigating avian influenza occurrences. The United States and these international partners have led global efforts to encourage countries to heighten surveillance for outbreaks in poultry and significant numbers of deaths in migratory birds and to rapidly introduce containment measures. The U.S. Agency for International Development (USAID) and the U.S. Department of State, the U.S. Department of Health and Human Services (HHS), and Agriculture (USDA) are coordinating future international response measures on behalf of the White House with departments and agencies across the federal government".

Together steps are being taken to "minimise the risk of further spread in animal populations", "reduce the risk of human infections", and "further support pandemic planning and preparedness".

Ongoing detailed mutually coordinated onsite surveillance and analysis of human and animal H5N1 avian flu outbreaks are being conducted and reported by the USGS National Wildlife Health Centre, the Centres for Disease Control and Prevention, the World Health Organisation, the European Commission, and others.

Treatment

There is no highly effective treatment for H5N1 flu, but oseltamivir (commercially marketed by Roche as Tamiflu), can sometimes inhibit the influenza virus from spreading inside the user's body. This drug has become a focus for some governments and organisations trying to prepare for a possible H5N1 pandemic. On April 20, 2006, Roche AG announced that a stockpile of three million treatment courses of Tamiflu are waiting at the disposal of the World Health Organisation to be used in case of a flu pandemic; separately Roche donated two million courses to the WHO for use in developing nations that may be affected by such a pandemic but lack the ability to purchase large quantities of the drug.

However, who Expert Hassan Al-Bushra has said:

> *"Even now, we remain unsure about Tamiflu's real effectiveness. As for a vaccine, work cannot start on it until the emergence of a new virus, and we predict it would take six to nine months to develop it. For the moment, we cannot by any means count on a potential vaccine to prevent the spread of a contagious influenza virus, whose various precedents in the past 90 years have been highly pathogenic".*

Animal and lab studies suggest that Relenza (zanamivir), which is in the same class of drugs as Tamiflu, may also be effective against H5N1. In a study performed on mice in 2000, "zanamivir was shown to be efficacious in treating avian influenza viruses H9N2, H6N1, and

H5N1 transmissible to mammals". In addition, mice studies suggest the combination of zanamivir, celecoxib and mesalazine looks promising producing a 50% survival rate compared to no survival in the placebo arm. While no one knows if zanamivir will be useful or not on a yet to exist pandemic strain of H5N1, it might be useful to stockpile zanamivir as well as oseltamivir in the event of an H5N1 influenza pandemic. Neither oseltamivir nor zanamivir can currently be manufactured in quantities that would be meaningful once efficient human transmission starts. In September, 2006, a WHO scientist announced that studies had confirmed cases of H5N1 strains resistant to Tamiflu and Amantadine. Tamiflu-resistant strains have also appeared in the EU, which remain sensitive to Relenza.

Epidemiology

The earliest infections of humans by H5N1 coincided with an epizootic (an epidemic in nonhumans) of H5N1 influenza in Hong Kong's poultry population. This panzootic (a disease affecting animals of many species, especially over a wide area) outbreak was stopped by the killing of the entire domestic poultry population within the territory. However, the disease has continued to spread. On December 21, 2009 the WHO announced a total of 447 cases which resulted in the deaths of 263.

Contagiousness

H5N1 is easily transmissible between birds facilitating a potential global spread of H5N1. While H5N1 undergoes mutation and reassortment, creating variations which can infect species not previously known to carry the virus, not all of these variant forms can infect humans. H5N1 as an avian virus preferentially binds to a type of galactose receptors that populate the avian respiratory tract from the nose to the lungs and are virtually absent in humans, occurring only in and around the alveoli, structures deep in the lungs where oxygen is passed to the blood. Therefore, the virus is not easily expelled by coughing and sneezing, the usual route of transmission.

H5N1 is mainly spread by domestic poultry, both through the movements of infected birds and poultry products and through the use of infected poultry manure as fertilizer or feed. Humans with H5N1 have typically caught it from chickens, which were in turn infected by other poultry or waterfowl. Migrating waterfowl (wild ducks, geese and swans) carry H5N1, often without becoming sick. Many species of birds and mammals can be infected with HPAI

A(H5N1), but the role of animals other than poultry and waterfowl as disease-spreading hosts is unknown. According to a report by the World Health Organisation, H5N1 may be spread indirectly. The report stated that the virus may sometimes stick to surfaces or get kicked up in fertilizer dust to infect people.

Virulence

H5N1 has mutated into a variety of strains with differing pathogenic profiles, some pathogenic to one species but not others, some pathogenic to multiple species. Each specific known genetic variation is traceable to a virus isolate of a specific case of infection. Through antigenic drift, H5N1 has mutated into dozens of highly pathogenic varieties divided into genetic clades which are known from specific isolates, but all currently belonging to genotype Z of avian influenza virus H5N1, now the dominant genotype.

H5N1 isolates found in Hong Kong in 1997 and 2001 were not consistently transmitted efficiently among birds and did not cause significant disease in these animals. In 2002 new isolates of H5N1 were appearing within the bird population of Hong Kong. These new isolates caused acute disease, including severe neurological dysfunction and death in ducks.

This was the first reported case of lethal influenza virus infection in wild aquatic birds since 1961. Genotype Z emerged in 2002 through reassortment from earlier highly pathogenic genotypes of H5N1 that first infected birds in China in 1996, and first infected humans in Hong Kong in 1997. Genotype Z is endemic in birds in Southeast Asia, has created at least two clades that can infect humans, and is spreading across the globe in bird populations. Mutations are occurring within this genotype that are increasing their pathogenicity. Birds are also able to shed the virus for longer periods of time before their death, increasing the transmissibility of the virus.

Transmission and Host Range

Infected birds transmit H5N1 through their saliva, nasal secretions, feces and blood. Other animals may become infected with the virus through direct contact with these bodily fluids or through contact with surfaces contaminated with them. H5N1 remains infectious after over 30 days at 0 °C (32.0 °F) (over one month at freezing temperature) or 6 days at 37 °C (98.6 °F) (one week at human body temperature) at ordinary temperatures it lasts in the environment for weeks. In Arctic temperatures, it doesn't degrade at all.

Because migratory birds are among the carriers of the highly pathogenic H5N1 virus, it is spreading to all parts of the world. H5N1 is different from all previously known highly pathogenic avian flu viruses in its ability to be spread by animals other than poultry.

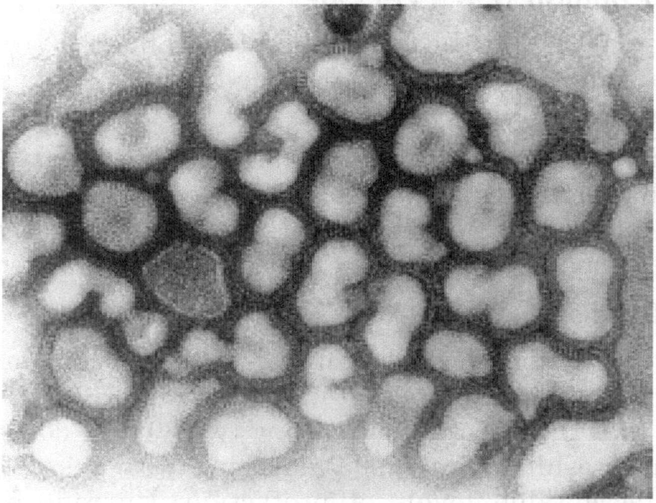

***Figure** : Influenza A virus, the virus that causes Avian flu. Transmission electron micrograph of negatively stained virus particles in late passage.*

In October 2004, researchers discovered that H5N1 is far more dangerous than was previously believed. Waterfowl were revealed to be directly spreading the highly pathogenic strain of H5N1 to chickens, crows, pigeons, and other birds, and the virus was increasing its ability to infect mammals as well. From this point on, avian flu experts increasingly referred to containment as a strategy that can delay, but not ultimately prevent, a future avian flu pandemic.

"Since 1997, studies of influenza A (H5N1) indicate that these viruses continue to evolve, with changes in antigenicity and internal gene constellations; an expanded host range in avian species and the ability to infect felids; enhanced pathogenicity in experimentally infected mice and ferrets, in which they cause systemic infections; and increased environmental stability."

The New York Times, in an article on transmission of H5N1 through smuggled birds, reports Wade Hagemeijer of Wetlands International stating, "We believe it is spread by both bird migration and trade, but that trade, particularly illegal trade, is more important". On September 27, 2007 researchers reported that the H5N1 bird flu virus can also pass through a pregnant woman's placenta to infect the fetus. They also found evidence of what doctors had long suspected—that the virus not only affects the lungs, but also passes throughout the body into the gastrointestinal tract, the brain, liver, and blood cells.

Approved Human H5N1 Vaccines

On April 17, 2007 the US FDA approved "Influenza Virus Vaccine, H5N1" by manufacturer Sanofi Pasteur Inc for manufacture at its Swiftwater, PA facility.

In March of 2006, Hungarian Prime Minister Ferenc Gyurcsany reported that Omninvest developed a vaccine to protect humans against the H5N1 influenza strain. The vaccine was approved by the country's national pharmaceutical institute for commercial production.

Results of Trials

Early results from H5N1 clinical trials showed poor immunogenicity compared to the 15 mcg dose that induces immunity in a seasonal flu vaccine. Trials in 2006 and 2007 using two 30 mcg doses produced unacceptable results while a 2006 trial using two doses of 90 mcg each achieved acceptable levels of protection. Current flu vaccine manufacturing plants can not produce enough pandemic flu vaccine at this high dose level.

"Adjuvanted vaccines appear to hold the greatest promise for solving the grave supply-demand imbalance in pandemic influenza vaccine development. They come with obstacles—immunologic, regulatory, and commercial—but they also have generated more excitement than any other type of vaccine thus far. In August 2007, scientists working with a GlaxoSmithKline formula published a trial of a two-dose regimen of an inactivated split-virus vaccine adjuvanted with a proprietary oil-in-water emulsion; after the second injection, even the lowest dose of 3.8 mcg exceeded EU criteria for immune response. And in September, Sanofi Pasteur reported in a press release that an inactivated vaccine adjuvanted with the company's own proprietary formula induced EU-accepted levels of protection after two doses of 1.9 mcg."

The "GlaxoSmithKline-backed team that described an acceptable immune response after two adjuvanted 3.8-microgram (mcg) doses found that three fourths of their subjects were protected not only against the clade 1 Vietnam virus on which the vaccine was based, but against a drifted clade 2 virus from Indonesia as well. To achieve prepandemic vaccines, researchers would have to ascertain the right dose and dose interval, determine how long priming lasts, and solve the puzzle of measuring primed immunity. Further, regulatory authorities would have to determine the trial design that could deliver those answers, the public discussion that would be necessary for prepandemic vaccines to be accepted, and the safety data that would need to be gathered once the vaccines went into use".

Individual Studies

Revaccination - January 2006

The purpose of this study is to determine whether having received an H5 vaccine in the past primes the immune system to respond rapidly to another dose of H5 vaccine. Subjects who participate in this study will have participated in a previous vaccine study (involving the A/Hong/Kong/97 virus) during the fall of 1998 at the University of Rochester.

A/H5N1 in Adult - February 2006

The purpose of this study is to determine the dose-related safety of flu vaccine in healthy adults. To determine the dose-related effectiveness of flu vaccine in healthy adults approximately 1 month following receipt of 2 doses of vaccine. To provide information for the selection of the best dose levels for further studies.

H5 Booster after Two doses - June 2006

The purpose of this study is to determine whether a third dose of vaccines containing A/Vietnam/1203/04 provides more immunity than two doses. Subjects who participate in this study, will have participated in DMID protocol 04-063 involving the A/Vietnam/1203/04. In this study, each subject will be asked to receive a third dose of the H5 vaccine at the same level administered in protocol 04-063.

H5 in the Elderly - August 2006

This study is intended to examine the safety and dose-related immunogenicity of three dosage levels of the Influenza A/H5N1 vaccine, as compared to saline placebo, given intramuscularly to healthy elderly adults approximately 4 weeks apart.

H5 in Healthy Adults - November 2006

This randomised, controlled, double-blinded, dose-ranging, Phase I-II study in 600 healthy adults, 18 to 49 years old, is designed to investigate the safety, reactogenicity, and dose-related immunogenicity of an investigational inactivated influenza A/H5N1 virus vaccine when given alone or combined with aluminum hydroxide. A secondary goal is to guide selection of vaccine dosage levels for expanded Phase II trials based on reactogenicity and immunogenicity profiles. This dose optimisation will be applied to both younger and older subject populations in subsequent studies. Subjects who meet the entry criteria for the study will be enrolled at one of up to 5 study sites and will be randomised into 8 groups to receive two doses of influenza A/H5N1 vaccine containing 3.75, 7.5, 15, or 45 mcg of HA with or without aluminum hydroxide adjuvant by IM injection (N= 60 or 120/vaccine dose group).

Bird flu - November 2006

This study is designed to gather critical information on the safety, tolerability, and the immunogenicity (capability of inducing an immune response) of A/H5N1 virus vaccine in healthy adults. Up to 280 healthy adults, aged 18 to 64, will participate in the study. Each subject will participate for 7 months and will be randomly placed in one of several different study groups receiving a different dose of vaccine, vaccine plus adjuvant, or placebo. All subjects will receive two injections of their assigned study product, about 28 days apart, in their muscle tissue. Subjects will keep a journal of their temperature and any adverse effects between study visits. A small amount of blood will also be drawn before the first injection, 7 days after each injection, and 6 months after the second injection.

Pandemic Flu - January 2007

This Australian study will test the safety and immunogenicity of an H5N1 pandemic influenza vaccine in healthy adults.

Children - February 2007

This is a randomised, double-blinded, placebo-controlled, staged, dose-ranging, Phase I/II study to evaluate the safety, reactogenicity, and immunogenicity of 2 doses of an IM inactivated influenza A/H5N1 vaccine in healthy children, aged 2 through 9 years. This study is designed to investigate the safety, tolerability, and dose-related immunogenicity of an investigational inactivated influenza A/H5N1 vaccine. A secondary goal is to identify an optimal dosage level of the vaccine that generates an acceptable immunogenic response, while maintaining an adequate safety profile.

Bibliography

Abbas, M.A.: *Effect of Organic Amendments and EM on Crop Production in Pakistan,* SP, Brazil. Pub. USDA. Washington, D.C., 19991.

Ahn., D.U. : *Effects of Post-Mortem Time Before Chilling and Chilling Temperatures on Water Holding Capacity and Texture of Turkey Breast Muscle*, Poultry Science, 1997.

Blackbourn, David: *The Long Nineteenth Century: A History of Germany,* Oxford University Press, New York, 1998.

Boorman, K.N.: *Poultry Feedstuffs, Supply Composition and Nutritive Value,* CABI Publishing, UK, 2002.

Charles, T. Burr and Homer O. Stuart: *Commercial Poultry Farming*, Biotech Books, Delhi, 2011.

Cotterill, O.J.: *Egg Science and Technology,* Food Products Press, Imprint of Haworth Press, New York, London, 1995.

Donald, E. Green: *A History of the Oklahoma State University Division of Agriculture,* Oklahoma State University, Stillwater, 1990.

Elson, H.A.: *Poultry Production Systems, Behaviour, Management and Welfare,* CAB International, NY, 1992.

Ferguson, M.W.J.: *Egg Incubation, Its Effects on Embryonic Development in Birds and Reptiles,* Cambridge University Press, UK, 1991.

Grist, A.: *Poultry Inspection, Anatomy, Physiology and Disease Conditions,* Nottingham University Press, UK, 2006.

Hambidge, Gove : *Diseases and Parasites of Poultry*, Biotech Books, Delhi, 2004.

Hernandez, J.: *Optimum Egg Quality: A Practical Approach,* 5M Publishing, U.K, 2007.

Homer O. Stuart: *Commercial Poultry Farming*, Biotech Books, Delhi, 2011.

Jordan, F & Pattison, M & Alexander, D & Faragher, T.: *Poultry Diseases,* WB Saunders, UK, 2002.

Keith Wilson N.D.P: *A Handbook of Poultry Practice*, Agrobios, Delhi, 2000.

Lakhotia, R.L. : *Agro's Dictionary of Poultry Science*, Agrobios, Delhi, 2006.

Larbier, M.; Leclercq, B.: *Nutrition and Feeding of Poultry*, Nottingham University Press, U.K., 1994.

Mandal, A.B. : *Nutrition and Disease Management of Poultry*, International Book Distributing Co, Delhi, 2004.

McLelland, J.: *A Colour Atlas of Avian Anatomy,* Wolfe Publishing, London, 1990.

Mudd, D.: *Profitable Poultry: Raising Birds on Pasture*, USDA's Sustainable Agriculture Network (SAN), Washington, DC, 2001.

Nandi, S. and S. Samanta: *Poultry Diseases: At a Glance*, IBDC Pub, Delhi, 2010.

Nicholls, C.: *The Workboot Series: The Story of Eggs,* Kondinin Group, Cloverdale W.A., 2005.

Nowland, W.J.: *Modern Poultry Management in Australia,* Rigby, 1978.

Owen, W Powell : *Poultry Farming and Keeping*, Biotech Books, Delhi, 2005.

Panda, A K ; S V Rama Rao and M R Reddy: *Growth Promoters in Poultry : Novel Concepts*, International Book Dist, Delhi, 2008.

Powell., C. : *Microbiological and Hydraulic Evaluation of Immersion Chilling for Poultry*, Journal of Food Protection, 1995.

Rai, Manoj Kumar : *Handbook Of Poultry*, Oxford Book Company, Delhi, 2011.

Randall, C.J.: *Color Atlas of Diseases and Disorders of the Domestic Fowl and Turkey,* Iowa State University Press, UK, 1991.

Sainsbury, D.: *Poultry Health and Management,* Blackwell Science, US, 2000.

Sharma, R. N.: *Poultry Management*, Vista International Publishing House, Delhi, 2008.

Tsushima, T. : *Survey of Bacterial Contimanation and Microbiological Control at a Poultry Slaughterhouse.* Journal of the Japan Veterinary Medical Association, 1996

Vegad, J L : *Poultry Diseases : A Guide for Farmers and Poultry Professionals*, International Book Dist, Delhi, 2008.

Wages, Dennis: *Biosecurity in the Poultry Industry*, International Book, Delhi, 2006.

Whitehead, C.C.: *Bone Biology and Skeletal Disorders in Poultry,* Carfax Publishing Company, U.K., 1992.

Yamamoto, T. & Juneja, L.R. & Hatta, H.: *Hen Eggs: Basic and Applied Science,* CRC Press, Delhi, 1996.

Young, M.: *Controlling Newcastle Disease in Village Chickens,* ACIAR, US, 2002.

Index

❑❑❑